The Art of DECORATING WITH FLOWERS

The Art of DECORATING WITH FLOWERS

Pamela Westland

Ward Lock Limited · London

ISBN 0 7063 5027 8

First published in Great Britain 1976
by Ward Lock Limited,
116 Baker Street, London, W1M 2BB

Line drawings by Delia Delderfield
Designed by Paula Hawkins
House editor Dorothy Ward

Text filmset in Apollo

Printed and bound by
Cox & Wyman Ltd,
London, Fakenham and Reading

Contents

Chapter 1

Ready to Start

WHENEVER I buy a practical book I want to get started on the ideas straight away; copy a design, follow a recipe, achieve something however small within minutes. As soon as I have done this, I feel confidence in the rest of the book.

With a book on flower decorating, this is quite simple to do, even if it is in the middle of winter, pouring with rain and you cannot get outside the door.

There are so many natural materials, everyday things we all tend to collect, that have a perfect affinity with flowers; it is quite fun at first just to gather them all together and look at them with new eyes. For just as beauty advisers say that to 'think thin' is to be half-way to a successful diet, so it is with designing. To start thinking in terms of the design possibilities in the materials all around us is to be well on the way to be bursting with ideas.

The designs in this book are not the stunning award-winning kind, where flowers are shown in accordance with a set of rules and with conventional formality. Quite the reverse, in fact. I have tried to show the flowers at their simplest and – I believe – most attractive. None of the designs requires complicated equipment, expensive accessories or special skill. They simply call for a love and appreciation of the flowers and other natural materials.

Take shells. Most households have a handful of shells somewhere, relics of a sea-side vacation or excursion, treasured souvenirs yet often not quite enough in themselves to make anything worthwhile. Shells are perfect partners for flowers. Dish-shaped ones like the ridged scallops or craggy oysters make perfect little containers, like bowls of pure pearl, to hold fresh flowers, and so if you can snatch even a handful of flowers or weeds you can make a start. (One of the designs in *Chapter 8*, on *page 100*, shows a study in green, grouped together in a trio of shells.) And if you have a few dried flowers, maybe salvaged from another arrangement another time, see how the translucence of the shell interior shines a fresh light on them; no other container could possibly be more flattering.

Hollow shells, such as tritons, can be planted with trailing greenness or a clutch of colourful little alpines, and make perfect miniature indoor gardens. If you have a few small houseplants do a spot of transplanting into some shell *jardinières*, just to get the feel of things.

And shells of every describable kind can be combined with dried flowers and leaves as a decorative way to frame pictures, prints and photographs – especially sea-side ones. They give the artwork a new dimension.

Then there are cones. They are delightful in themselves, each one like an intricate and exquisite piece of woodcarving. Tiny ones nestling among the needles of conifer trees can be preserved on the branches for all time in glycerine – the needles will turn nut brown, too. But you probably will not have an armful of conifer branches to hand. What you might have to start with is a bag of cones, beech nuts and mast, acorns and cups, seedpods, keys, seedheads and such things, the hoard from many a woodland walk. All you need besides is a piece of fabric or card in a sympathetic colour for the background, plus a tube of glue, then you can start creating – a picture that will be a three-dimensional study in moody browns.

All these materials we can pick up freely, for they are the gifts that nature has left us, as if we were on a treasure hunt and these were the prizes. But flowers are different. We have to have a very special sense of responsibility about picking them, particularly now when so many genera are threatened by extinction. Some of the designs here show wild flowers which grow in profusion in some parts of the world, but not everywhere. If you think of using wild flowers for decoration, never pick one which you know is rare or on a conservation list. Substitute another, similar in colour and shape, for there cannot be any pleasure in creating a design which leaves the countryside less beautiful for others to enjoy. None of the arrangements throughout the book relies on using any particular flower only. If a photograph shows a flower which has long, blue spikes, look for one with the same characteristics, wild or cultivated; if there is a mass of tiny white blooms chosen to give a fluffy effect, see what you can find which serves the same purpose. It is the overall effect that counts.

Even picking cultivated flowers needs thought. Unless you grow a bed or rows of flowers especially for cutting, you do not want to deprive the garden in its moment of glory. This point was effectively brought home to me by an old lady I once knew. She lived in a huge house, completely hidden by dark trees and surrounded by woodlands stretching as far as the eyes could see. Early in the springtime these woods would be carpeted with tiny wild flowers, each one as pretty as a bride's headdress. It was a breathtaking sight, such a profusion of flowers, but even so the old lady had a very strict code of practice. 'Let's play a game,' she would say to visiting children. 'You may pick as many flowers as you like while I put the kettle on, then I will come out. If I can see where you have taken the flowers from, you'll get no tea!' And since tea with her always included the darkest chocolate cake and crispest buns ever, that was

Opposite, before they are arranged, and even on the journey home, flowers benefit from a long, cool drink. Details on page 15
Next page, a dry, airy garage or garden shed is an ideal place to hang bunches of drying flowers and grasses. Details on page 26

incentive enough to pick only a few from each patch. She never would allow anyone to dig up a bulb, either, but encouraged them instead to visit her again when the flowers were over, and collect the seed. She would have approved of the notes in *Chapter 9*.

Capturing the mood

All flowers, both wild and cultivated, have such individual characteristics; such differences in size, shape, colour, form; how they grow; where they grow; when they grow, that in dealing with them we are constantly faced with their various personalities. It seems to me, therefore, that the art of decorating with flowers lies in meeting the challenge to interpret these varying life-styles in one's own particular way yet always, of course, in a way which flatters the flowers. So whenever I see flowers growing, anywhere, I find myself thinking how they would look indoors, the kind of backgrounds that would be most suitable, the different moods they could create and the type of container that would be best.

When we see flowers growing by themselves, a single type stealing the limelight, I think this is a clue to how we should display them. Accordingly, there are several designs in the following chapters where only one type of flower is used, arranged as naturally as possible. There are two designs with poppies, both evocative of the way they grow in the countryside or in our gardens. One is as casual as can be, with the flowers in a pottery jar, vibrant against a handful of pale green summer grasses, and the other more lasting, with a background created of dried cereals, grasses and seedheads so that the poppies, against a natural wood background, are projected as vividly as they are in a field.

There are two arrangements of tree lupins, too, both recognizing that these flowers have a strong mind of their own, with stems that twist and turn and defy formality and precision. The flowers are shown in an iron charcoal burner, the kind of thing one might take on beach picnics where, indeed, the flowers can be found growing, and in a different mood, curling around drift-wood, again reminiscent of sand, sea and breezes.

Just occasionally, it is possible to try to improve on nature and arrange flowers to show them more beautifully that they appear as they grow. This is possible with members of the wild cabbage family, like mustard and charlock which, seen from a distance, present a carpet of bright golden yellow. Yet, close up, each plant has no more than a few straggly blooms at the end of watery stems. If you concentrate on the flowers, and cut the stems down to size, you can restore the importance of the flowers and bring back the apparent intensity of colour. On *page 87* a design combining charlock and mustard in a honey jar, shows how effective this treatment can be.

I wouldn't say that any design could actually improve on the breathtaking beauty of a hawthorn tree in full blossom, yet it is amusing and interesting to show a few small twigs close up, to emphasize the minute scale of the confetti-like flowers. In a tiny container mixed with just one other flower type you can create an amusing change of roles – even lawn daisies look as large as chrysanthemums by comparison!

It is this awareness of the way that they grow that, for most of us, is the secret of decorating with wild flowers.

Previous page, small, flat flower heads with shortened stalks are about to be covered with more ground silica gel. Details on page 28
Opposite, nuts, cones and berries, shiny or spiky, can be preserved on the spray with leaves, in glycerine solution. Details on page 30

Chapter 2

Fresh as a Daisy

CUTTING flowers for the house is a way of borrowing, if only for a few days, a little of the natural beauties of garden and countryside; of recreating the casual, carefree feeling of the flowers as they grow.

It is quite a big step for flowers to take, from growing unhindered in a sheltered border, on a grassy bank or under a shady tree, to being suddenly imprisoned in a container in a well-heated room. In other chapters we examine the various processes that enable us to dry, preserve or press flowers for later use in designs. Here we are concerned with those things we can do both to make the transition from outdoors to indoors as easy as possible and to help the flowers last well in water.

Beginning at the beginning, it is important just how and when the flowers are cut. Always use good secateurs, scissors or a very sharp knife, and (in most cases) cut the stems firmly and cleanly across at an angle. Never pick the flowers with your fingers. This is liable to damage both stem and plant.

Next, the time of day. Unless it is really unavoidable, never cut flowers in the heat of the sun; this is when they are at their most vulnerable and are most likely to wilt. Early in the morning, late afternoon or evening are best.

Then, the stage of development. Flowers cut at a very immature stage will not develop properly in water – roses are an example of this. The buds will simply hang their heads and die. Equally, flowers left on the plant until their centres are fluffy with golden pollen will not be very rewarding. They are already on the downward slope of the development graph. To enjoy the maximum benefit from the flowers, cut them when they are just about to come into full bloom. Daffodils are exceptions, as everybody knows; they will burst forth quite cheerfully in water from tightly furled buds into full trumpet flowers.

Long spikes of flowers, such as lupin, delphinium or larkspur, are best cut when the lower flowers are fully open and the top buds just about to burst. As the lower ones wilt you can snip them off, leaving those higher up the stem to take over.

The kindest thing you can do for cut flowers is to give them a good long drink of water at once. Take a bucket of water round the garden with you and plunge the stems in as you cut them. Even gathering wild flowers in the country, it might still be possible to stand them in water for an hour. The colour photograph on *page 9*, taken on a sailing boat shows a sheaf of wild flowers in jam jars with chunky, improvised string handles, taking refreshment before being arranged.

If flowers must travel after cutting it is important to have them well wrapped up. Use large plastic bags lined with dampened tissues or newspaper and pack the flowers loosely. Blow up the bags like balloons and secure them tightly at the top. Carry them in a covered basket or container if possible, but anyway protected from the sun.

As soon as you get the flowers home, stand them deep in water, preferably overnight. Contrary to belief, rainwater from a barrel or tank is not best for the purpose as it can contain too many bacteria. But icy-cold water straight from the tap can be rather severe, so draw it off before you go out and leave it to stand. Put the flowers in a cool, dark place while they are reviving and make sure they will not be exposed to the full glare of the sun.

Remove all the leaves that will come below the water line when the flowers are arranged. They not only cause the absorption of an unnecessary amount of water, but taint and discolour the water left in the container. If possible, however, allow a few of the leaves higher up the stem to remain, so that the natural plant process of photosynthesis can continue.

Conditioning

All flowers benefit from being stood in water as soon as they have been picked and some need further conditioning before being arranged. Although this might sound like rather too much trouble, consider the profits you will reap in terms of a long-lasting arrangement in the house. For the purposes of conditioning, the plant material is categorized according to the kind of stem it bears – woody, milky, hollow or soft.

Woody stems include all the trees and shrubs, besides flowers as varied as mallow, centaurea and thistles. Strip off any bark for 1 in from the cut, then lightly crush that part of the stem to break down the fibres and assist the intake of water. To do this, you need an improvised hammer and anvil. Use a wooden mallet (the kind used for banging in tent pegs), a wooden rolling-pin or a small hammer with a rounded end, and for the 'anvil' a stout block of wood or an old chopping or pastry board on a firm surface. Tap the stems with a gentle, rhythmic knocking action – not a mighty blow that will smash them.

The rhododendrons in the colour photograph on *page 54* were treated in this way. The arrangement shows an easy way to make the most of a few short-stemmed blooms, seen here floating in an old blue and white china bowl. Spray the flowers every day with cold water, using an old perfume atomizer or a clean garden sprayer. Needless to say (though I have sometimes forgotten to do so) remember to remove the arrangement to a stain-resistant surface before spraying, and wait until the water has finished dripping before you return it to a piece of polished furniture.

Milky stems have to be sealed at the ends to stop the sap escaping – for much the same reasons as one seals meat over a high heat before continuing to cook it. Poppies, fragile though they seem on the plant, will last in water for several days and members of the spurge family (*euphorbia*) for several weeks.

Stand the stems in water as soon as they are cut – about an hour is long enough. Then pat the stem ends dry with a tissue and burn them by dabbing them on to a red-hot piece of coal or charcoal, or holding them over a candle or match flame for a second or two. The stems will sizzle and seal across the cut. Put the flowers back in cold water until you are ready to arrange them. Then, as you cut them to the lengths you require, seal the ends again.

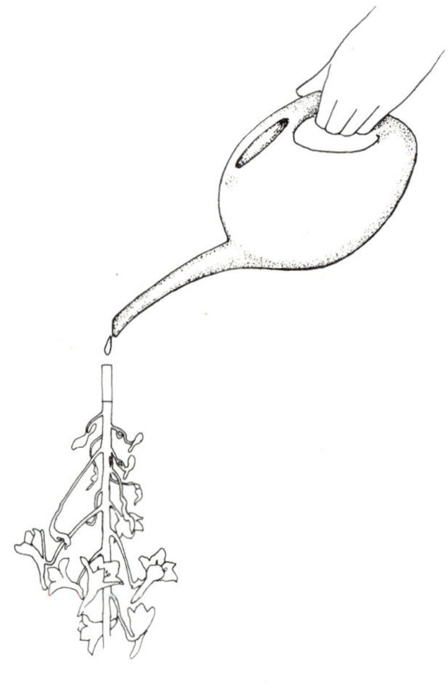

Figure A

Hollow-stemmed flowers benefit from a measure which, although it sounds rather drastic, is very simple to carry out. It is the method used to keep 'show' larkspur, delphinium and lupin spikes such triumphantly blazing towers of colour and is well worth the little trouble involved. Indeed, it is quite fun to do. Cut each stem straight across, not diagonally as usual, and hold it upside down. Using a medicinal dropper or a houseplant watering-can with a long, narrow spout, fill the stem with cold water (*figure A*). Tap it firmly with the side of your hand to make sure there are no air locks, then plug the end with a tightly twisted wad of dampened cotton wool (absorbent cotton). Stand the stems deep in cold water before arranging them.

The lupins in the colour photographs on *pages 44* and *53* were conditioned in this way and lasted in the arrangements for well over two weeks. It is interesting to note that other flowers, picked at the same time and not conditioned, drooped after only two days of standing in a bucket of water.

Soft stems such as hyacinth and daffodil should be recut under water to prevent air locks forming. Indeed, Japanese floral experts cut all stems under water. Have a wide, shallow bowl ready and use a pair of very sharp scissors or a sharp knife. Make the cuts cleanly at an angle across the stems and then stand them in cold water. Stems which have to be left out of water for any length of time will dry out and need to be recut under water again.

Many other stems which do not need to be scraped, crushed, burnt or filled with water benefit from being stood for a few minutes in a little boiling water. This applies to flowers such as campanula, hellebore, campion, primula, columbine, cinquefoil and potentilla. Cut the stems at an angle, pour about 1 in of boiling water into a heat-resistant container and stand the stems in it, making sure that all the ends are submerged. Roll a protective collar of tissue or newspaper and wrap it round the rim of the container to prevent the steam from reaching the flowers. Remove the stems and stand them in cold water before arranging.

The mechanics Obviously the most important equipment you need before you begin arranging flowers is a selection of containers. The day of the vase is past and containers can be anything from a soup ladle or tureen to a coffee mug or teapot, from a tin lid concealed behind a piece of wood to an iron barbecue. If you are using dried flowers, the range is endless because the container does not then need to hold water. Just looking through the colour photographs will give you plenty of ideas.

It is rarely possible to arrange flowers satisfactorily without any holding material at all; this can really only be done with a container which has a narrow enough neck to support the stems well. Otherwise, a block of artificial foam holding material is the easiest to work with. These blocks, sold under brand names such as Oasis, Florapak and Fil-fast may be found in all shapes and sizes from small cylinders and cubes suitable for miniature designs to blocks over 1 ft long. The largest can be used upright in large urns or on pedestals, or flat in bowls or dishes.

Judging the size of foam holding material you will need is rather like assessing the size of an iceberg. Calculate that you will need nearly as much foam above the rim of the container as there is below. Otherwise the design will lack flexibility and you will not be able to place stems to droop down over the rim. In a deep container, push the foam well down inside the neck so that it is firmly held in place. For large arrangements when you will be using heavy materials such as sprays of leaves or shrubs, it is advisable to 'wire' the foam. This means covering it with a piece of lightly crumpled wire netting, the kind with 1-in mesh, and pushing this firmly against the sides so that it grips well.

In a shallow container such as a plate or dish the foam will always need to be anchored. You can use pinholders specially made for the purpose, flat discs with spikes which hold the foam block. Test the rigidity of the foam. If it seems likely to topple, and particularly if your arrangement is to be asymmetrical, secure the holder to the container with a knob of extra tacky clay-like material, such as OasisFix florists' clay.

The principle of the artificial foam is that it holds water for a number of days – the actual time will, naturally, depend on the number and size of the flowers you use. To be fully effective it must be thoroughly soaked before using; it is not enough just to pour water over it. Place the block in a bowl of water and leave it for between half an hour and an hour, when it will sink to the bottom. If the block dries out before the flowers die, slowly and carefully pour water over it – but first, of course, remove the arrangement to a stain-resistant surface.

Foam can sometimes be re-used. This again depends on the volume of material it has held. Pull the stems straight out, at the angle they were pushed in, to avoid breaking the foam, drain the block thoroughly until dry and store in a plastic bag.

When only a few flowers are to be arranged, pinholders can be used to hold the stems in place. These are heavy metal discs with closely packed sharp vertical spikes. Again, in a shallow container, they should be secured with a piece of extra tacky florists' clay, and can be supplemented by a piece of wire netting over the top. I find that pinholders restrict a design more than foam material because the spikes are all at the same angle and it is not quite so easy to get 'movement' into an arrangement.

Ordinary glass marbles are another useful form of holding material, particularly attractive in a glass container where the mechanics show. A handful of marbles in a glass dish of, say, narcissus, becomes an important part of the design. Marbles of various sizes can be bought in bags.

It goes without saying that it is important to keep all containers, holding materials and other aids as clean as your cups and saucers. Tempting though it might be to put away a container complete with holding material 'ready for next time', it is just not good flower management!

Breakfast trio

First thing in the morning is not, perhaps, the best time for arranging flowers delicately, nor yet to expect compliments for our floral artistry. But however rushed the breakfast scene, any day will start more pleasantly for just a glimpse of a few fresh blooms, no matter how hastily arranged.

The three designs in the photograph opposite have been worked without any holding materials – who wants to be waiting for blocks of foam to soak while watching the egg timer? The flowers and leaves were all snatched from a waste patch at the bottom of a city garden. The 'lift' comes in the containers – silver-plate salt and mustard pots which, appropriately, hark back to the days when the first meal of the day was an altogether grander and more leisurely affair than today's single deep brown egg. Any of these designs would take its place delightfully on a breakfast table or tray, or on a bedside table in a guest bedroom.

The most colourful of the designs, in the large salt pot, is a marriage of clusters of minute flowers cut from a large tree, and small flowers – lawn daisies – made to look big and important by comparison.

Because a design of this kind will almost always be viewed from above, the focal point is the central cluster of red hawthorn flowers, snipped down into tiny bunches. The stems are lightly crushed by gentle tapping with a hammer before arranging so that they can more readily take up water. Seen against sprays of such tiny flowers the simple lawn daisies look as big as chrysanthemums. We carefully picked out ones which had pink-tinged petals to echo the colour of the hawthorn. At each side of the pot there is a creamy-white cluster of rowan (mountain ash) flowers – from a tree which is more usually beloved for its coral berries. The flowers repeat the shape of the hawthorn but the colour of the daisies. For complete colour contrast sprigs of forget-me-not are slipped in here and there. The leaves are hedge parsley, soft and pale green, and blackberry, placed to curve low towards the legs of the container and to provide a deep and mysterious background against which the daisies can be seen, petal by petal.

In the small salt pot, on the right, there is nothing more than a handful of mauve dead-nettle, the stalks necessarily cut short and so featuring only the top-most leaves, at their most fresh and youthful. The flowers are tiny statements in two shades of mauvey-pink, perfect with the glint of silver.

On the left, for the small mustard pot, another tall plant has been cut to the size required – and the flowers take on a new personality. These are garlic mustard, snowy

white four-petalled flowers with deep yellow centres and leaves with well defined veins. The pretty effect of spring-like green and white, almost bridal with its shower of confetti look, shows how beautiful simplicity can be when it is planned.

Terrace table

From breakfast in the garden to drinks on the terrace, perhaps. Still a no-fuss, no time to spare arrangement, this time worked in foam. Two semi-circles of foam, soaked to capacity, stand on an old dinner plate and in the centre there is a small bowl filled with a block of the same material.

This type of design could be copied for any circular table, a dining table especially since it is shallow and discreet enough not to draw the attention of the guests from each other! Blue and yellow are prettiest together when one is a sharp and one a muted colour – in this case the zinging yellow of wild charlock shines against the modest blue of the bluebells. Choose any other blue flowers with the same kind of flower and stem formation – grape hyacinths, hyacinths, harebells or larkspur.

Coming from the sturdy cabbage family, charlock or any of the wild mustards have good strong stems. But our blue flowers had fleshy pliable stems and needed the help of holes pushed in the foam with a knitting needle if they were not to break off shorter than was intended.

Opposite, single flowers, small sprays and leaves are pressed in blotting paper between the pages of a heavy book. Details on page 40

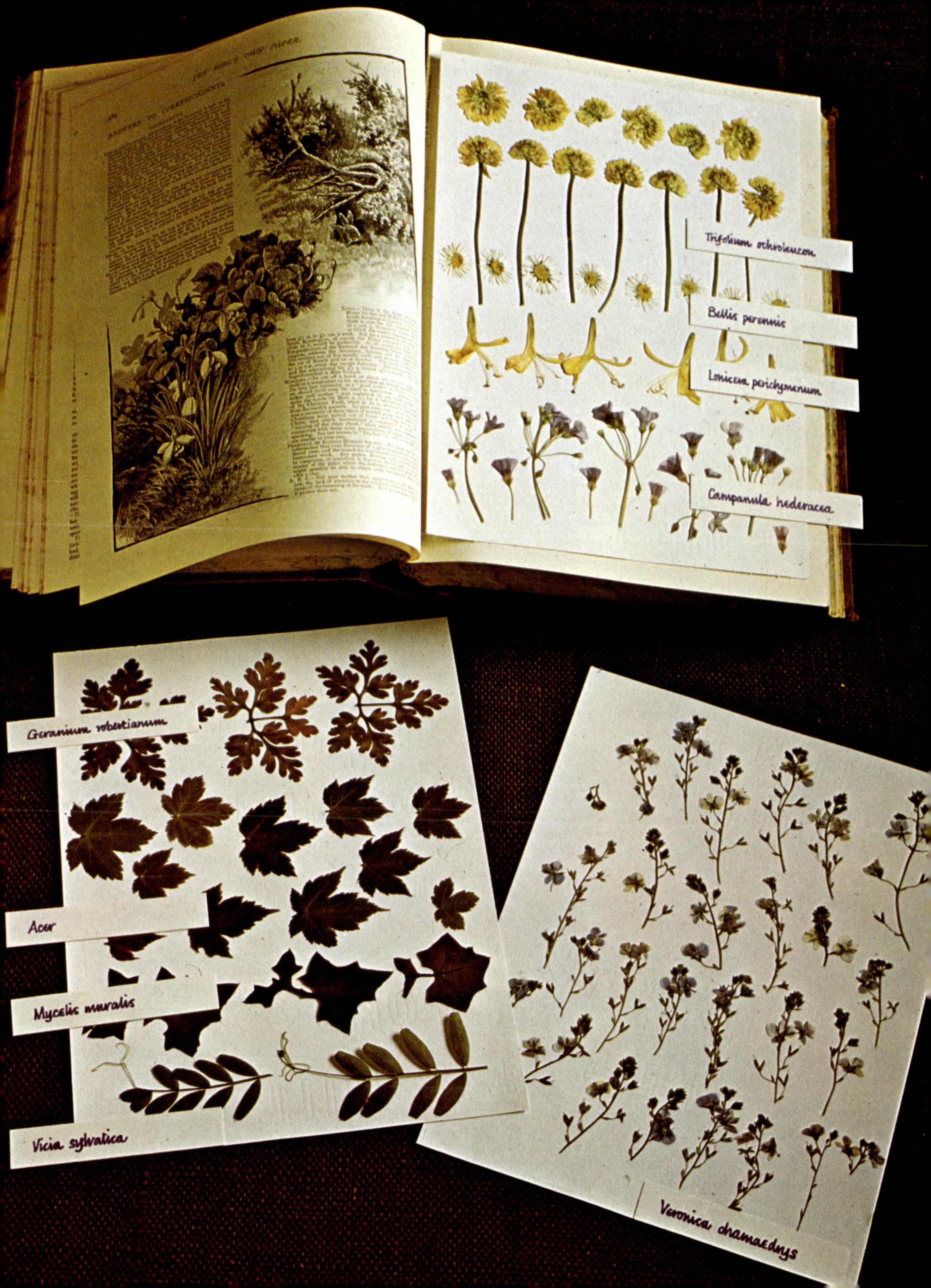

Trifolium ochroleucon
Bellis perennis
Lonicera periclymenum
Campanula hederacea
Geranium robertianum
Acer
Mycelis muralis
Vicia sylvatica
Veronica chamaedrys

To introduce another shape and separate the two strong colours, we chose an inner and then an outer ring of deep lustrous ivy leaves. As it happens, these repeat one of the motifs on the wrought iron table, but blackcurrant or young maple leaves, for example, would look just as well.

A bunch of daffodils

Sometimes it is the flowers which mean the most to us – the ones a child brings home in a bunch, those that represent the end of a long winter, or simply a few bought when there's not much money to spare, that present the most difficulty in arranging.

Opposite, herbs at a half-way stage, a decorative arrangement in a kitchen jug before they are used for cooking. Details on page 46

A bunch of daffodils is a case in point. If, like so many people who love flowers, you feel a tinge of guilt about just taking off the rubber band and lowering them into a vase of water, here's a way to show them in all their golden glory. The design shown in the black and white photograph on *page 23* features ten daffodils scattered among a mass of in-flower wild marjoram, which provides a deep, dark background for them, and intermingled with crisp, white sprays of lily of the valley and mostly cream anemones. The height and width of the design is set by the long curving sprays of soft, furry pussy willow.

Fill any wide-necked jug with well-soaked foam holding material, cutting the piece deep enough to push firmly into the container and extend well above the rim. Trim the branches of pussy willow so that the longest, for the centre back, will be about three times the height of the container and those at the sides a good deal shorter. Cut the twigs at a sharp angle and lightly crush the stems with a round-ended hammer. Stand them in water before arranging in the foam.

Cut the daffodils in graded heights, too, and position them tallest at the back, shortest at the front, with some facing to each side. It is important to arrange the daffodils into a well-rounded shape to give depth to the arrangement.

Now fill in the spaces between the daffodils with thick clumps of the flowering wild marjoram, or whatever you have that will serve the same purpose. In the centre, as a muted focal point, there is a single head of cow parsley (chervil), the purplish flowers sympathetic in appearance to the marjoram, yet clearer.

Break up the dark mass of the herb with sprays of lily of the valley. If you do not have any of these, improvise with something which offers a long, white shape – perhaps feathery stipa grass instead.

Then position the anemones, some at the back and sides of the design and others clustered over the rim of the jug, fulfilling a dual role: they must completely hide the green foam material.

The arrangement is complete now, except for a few sprays of leaves. Use these between the pussy willow to break up the hard lines and drooping low over the jug to blend with the decoration on the china.

In the same photograph you can see an old stone mustard pot decorated with pressed buttercup flowers. For details of this and other pressed flower design ideas, see *Chapters 7* and *10*.

Chapter 3

Collecting for the Future

COLLECTING plant material to dry or preserve for the future is rather like putting garden produce in the freezer; it will be there to draw on another time, another season, when the harvest is over and the gardens and fields are bare.

Nature, of course, does some of the drying process for us naturally, and a walk round the garden or in the country can produce armfuls of seedheads ready to dry, fallen leaves ready to press, all kinds of things ready to begin a new life as part of a decoration programme.

Start with grasses. They are a study in themselves and offer as much variety of shape and outline as flowers do. Admittedly they do not have the colour range, but then consider the part they play. They are the natural background for some of the most brilliant flowers around: bright red poppies would not look nearly as dramatic, bright blue cornflowers not nearly as blue if the colours of the surrounding grasses were fighting for equal rights. This is why grasses, in all their neutral shades, are so very useful in designs. They give the opportunity to have a great mound of fluffiness, a striking silhouette with more clearly defined features, or just a simple graceful curve, without upsetting or even necessarily adding to the colour balance.

Try to harvest grasses for drying before they begin to go to seed: in other words, while they are still thick with pollen. Once they start seeding they start disintegrating. At that stage, all you can do is to spray them with ordinary hair spray and hope for the best. Try, too, to cut grasses on a dry day, so that you do not bring in excess moisture in the form of dew or a recent shower. If it is a case of harvesting then or never, shake off the raindrops then gently bounce the grasses up and down on sheets of blotting paper.

Some of the lightest grasses can be pressed until they are as thin as paper and then used with flowers and leaves in pressed flower designs. More of this in *Chapter 4*. These and all other grasses can be dried hanging upside down; flat; or standing upright in a container, according to type and the head-to-stem weight ratio.

Hang bunches of the lightweight grasses upside down in a dry, airy place. It need not be too warm as the best results come with slow drying. Tie the bunches firmly with a natural material that has a little 'give', such as raffia or gardeners' twine, using a slip knot so that you can tighten it from time to time as the stems shrink. Hang the bunches over a clothes line, a clothes drying frame, on wire coathangers or from shelves or racks. Although a spare bedroom or storeroom are ideal, do not worry if the only space you have is in your living room. They actually look quite decorative, even at this stage.

Grasses with heavy heads should be dried flat, on trays or box lids, packed fairly loosely so that there can be a free circulation of air all round the heads, and turned occasionally. If you do not have a warm cupboard, the top of a wardrobe will do very well. (Just bear in mind that cats enjoy both the novelty of finding a new bed and the rustling noise that drying grasses make!) You can judge which grasses are candidates for this method by the size and weight. They include all those with round or oval fluffy-ball heads and those that hang in heavy clusters.

Grasses in the in-between category can be dried just by being stood upright in pots and jars, again loosely packed for the best results, or by either of the other methods.

Cereals like wheat or oats can be dried flat or upright. There seems little to choose between the two methods when the material has good, strong stems.

Wild oats, of course, are the enemy of the farmer and you will be doing him a good turn if you cut them – without damaging other crops. But wheat, oats, barley and rye are food crops and have a more important role in the scheme of things than finishing up decorating our homes. So be meticulous about cutting only from those plants which have sprung up on the edges of the fields, where the combine harvester could not reach them.

Missing out the flowers and leaves for a moment, since most of them need different treatment, we come to seedheads, a veritable bounty in themselves. It takes a trained eye to see at first glance the possibilities in these sometimes long, sometimes straggling spikes which we are accustomed to consigning to the compost heap or allowing to disintegrate on the plants. But in some cases – and this is as true of wild plants as of cultivated species – the seedheads are even more attractive and present more design opportunities than the flowers themselves.

When the bright blue flowers of the larkspur have long since gone, we are left with spikes of chubby little pods. Stripped of the longest lateral shoots, the spikes give welcome height to large arrangements, such as those on pedestals or intended to stand on the floor. Conversely, they can be snipped into literally dozens of separate components, each on a little stem, and used in miniature arrangements.

So it is with the tree lupins. If you miss them in flower you have a second crop to harvest, the stems of slightly furry pods which twist and turn and open into delightful wing shapes.

Mullein, some spikes standing as tall as a person, are another example. They dry with a thick mass of silvery-grey seedpods towering up like a pointed steeple – an impressive outline in any large-scale work.

All of these long spikes are best flat-dried on boxes. They should then be stored in buckets or other deep containers filled with clean, dry gravel or sand to hold the stems firm.

All the umbrella-shaped clusters of seedheads can be dried either by hanging or upright. These include yarrow in the *achillea* family, angelica, cow parsley, wild chervil and others of that type. Many of them are a soft lime green colour and so give a chance to introduce a new shade as well as texture and shape to a collection.

Many of the middle-height plants also offer new opportunities when the seedpods form. Mallow is a good example. The large pink, trumpet-shaped flowers transform themselves into star-like pods on stems, each one perfect for close work, where the heads can be used pushed flat into preformed foam shapes.

Poppy seedheads must be the best known of all those used in designs; the silvery-blue heads, shaped like classical urns, are indispensable. They can be dried by hanging upside down – but not before you have collected the minute black seed for next year's crop.

It would be tedious to extend the list of seedheads which can be dried because there are practically none that cannot. It is a question of looking at them with a new speculation, envisaging them in different surroundings and, in many cases, snipped from a single massive unit into a treasure of minute components.

Not many leaves can be dried satisfactorily by the no-fuss ways described. Indeed, wild carrot is one of the very few exceptions. The leaves can be tied in bunches and hung for about a week, when they will have retained most of their shape and taken on a few new curves. Other leaves tend to become so brittle that they are reduced to a fine powder as soon as they are touched. They respond far more readily to being preserved in glycerine, or dried by pressing. *See Chapter 4.*

There are precious few flowers, either, which can be dried unaided and, curiously, these do not seem to fall into any single category, nor share any obvious characteristics – except that, when hung in bunches to dry, they emerge not much less than a paler shadow of their former selves.

Blue comes top for colour retention by this method (though not by pressing). The deep blue of larkspur gives a tremendous lift to many of the designs to be worked out during the long winter months. And arrangers, if not all perfectionists, will forgive these long, wide branching spikes their very slight loss of lustre. The dried flowers can be used for large-scale designs in the full length of the stems or, for miniature designs, snipped into tiny stemlets or even individual florets. Cut the stems for drying when some of the top-most buds are still unopened. At this stage drying is most likely to succeed and the flowers emerge most attractive to look at. This is one of the flowers which makes such a valuable visual addition to pot-pourri, too. *See Chapter 5.*

Another blue success is viper's bugloss. Cut the stems when they have the widest range of fully developed and not-yet-developed flowers. Snip the leaves away all down the stems and hang only until the flowers are just dry. This is when they feel and sound papery. You do more harm than good by leaving them longer. The deep, trumpet shapes of the flowers should remain, the outlines broken by the long, pin-headed stamens.

Yellow has its successes, too. For beginner's luck, try hang-drying golden rod, another plant which offers material on two totally different scales – giant stems or miniature snippings, equally pretty in their way, with all the richness of a hoard of salvaged treasure. The daisy-like flowers of chamomile, so glowing when they are fresh, dry to an all-over soft beige, still pretty but not a mirror image of their youth.

Most pink flowers fade as they dry. The pale rose of sea pink or thrift, for example, becomes paler still, a sugar-almond-pink ball of flowers at the top of strong, straight stems.

The umbellifers are a group with a strong family likeness. We have already seen that the umbrella-shaped clusters of seedheads dry most attractively. To capture these plants at two stages of development, when they have two different personalities, cut some flowers for drying and leave some until the seed stage. The sometimes white, sometimes pink flowers are a great comfort to have in store; they do fill up space in large designs most efficiently.

The colour photograph on *page 10* shows a collection of flowers hanging to dry in a garage. They include some of the umbellifers, sea lavender, dock, oats, golden rod, water grass and various seedheads.

Drawing out the moisture

So much for drying flowers simply by hanging. The other way is to cover them completely with powder or crystals which will gradually draw the moisture from the petals. It works wonders for all but the most fragile of flowers, which would be crushed even by a powder carefully applied, and for the largest of sprays. This last limitation is, in fact, purely one of cost; the process would actually work quite satisfactorily but it would need a very large amount of drying crystals.

About 1 lb weight of drying crystals will dry about six to eight ox-eye daisies, perhaps a dozen chamomile flowers or four or five short sprays of vetch. The box in the colour photograph on *page 11* would be using about $1\frac{1}{2}$ lb of the crystals when the flowers were covered.

The two desiccants most frequently used are household borax powder (not the medical quality, which is more expensive) and ground crystals of silica gel. These crystals are dark blue. As they absorb the moisture from the flowers they turn light blue and later pale pink. To dry them out again for re-use simply spread the crystals out on a tray and put in a low-temperature oven or (on a not-too-windy day) in the sun. Store them in an airtight container for use over and over again.

There really is nothing to this method of drying flowers. For the container you can use any airtight box or tin – plastic sandwich boxes are ideal. Put a fine layer of the desiccant in the bottom, say $\frac{1}{4}$–$\frac{1}{2}$ in. Snip the stems of single flowers to no more than 2 in long and push them into the powder with the flower faces upright. Sprinkle some of the desiccant into your cupped hand. Support each flower in turn with the other hand and very gently let the powder run through your fingers until each and every petal is completely covered. Care now pays dividends later, and with a little practice you

Figure B

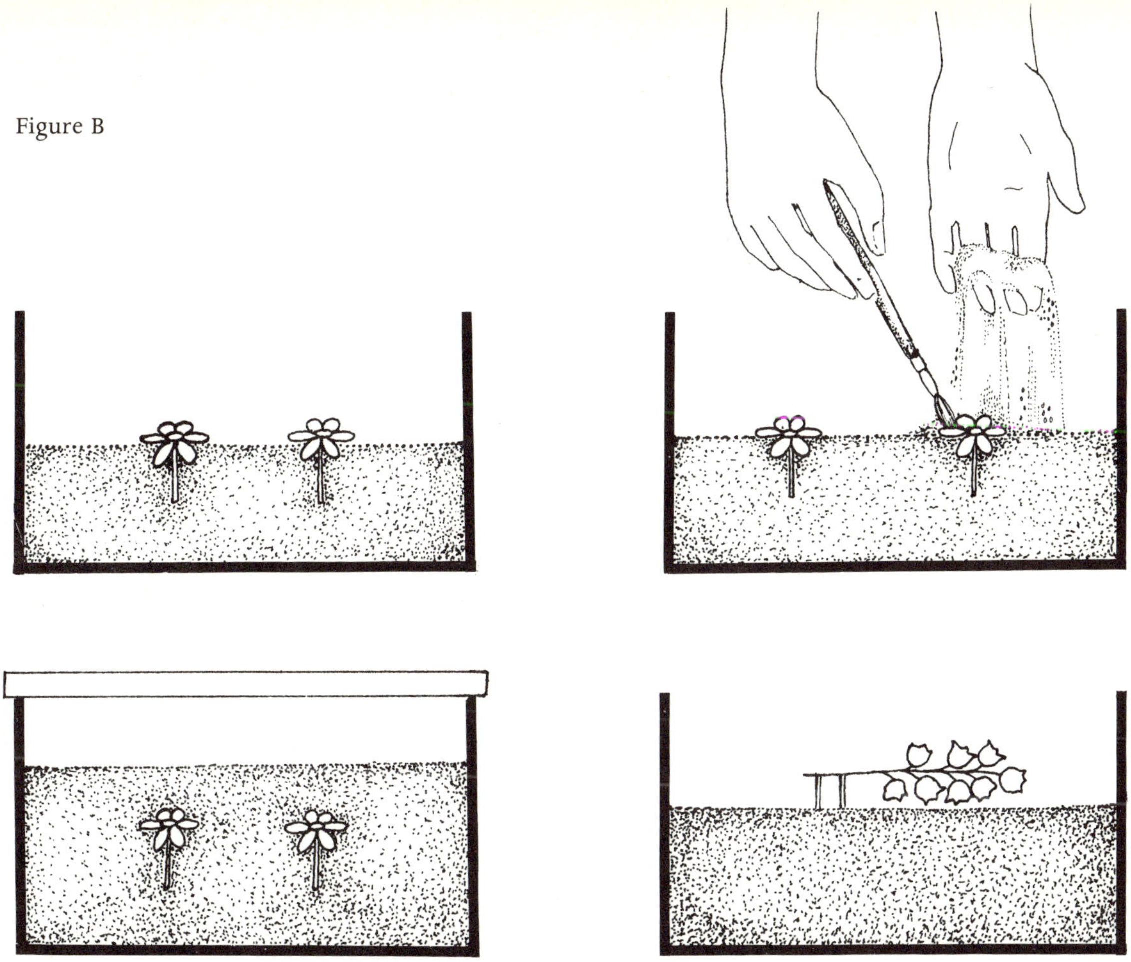

will soon get the feel of helping the flowers to retain their natural shape. It is the shape they take up at this stage that they will be when you remove them, so if you accidentally crush or bend back the petals with the drying agent it will not be their fault if they emerge slightly disfigured.

Cut sprays of vetch, wild pea and other similar types of flower into manageable lengths, say four or five inches, and lay these horizontally in the box. Make sure that the powder penetrates into the depth of each flower. Hold larger, trumpet-shaped flowers singly in your cupped hand. Pour the drying agent into the hollow, like pouring tea into a cup, until it is filled right to the top. Then gently lay each flower on the bed of desiccant in the box (*figure B*) and support it while you pour more of the material around it.

Cover the flowers with enough desiccant to enclose them completely, replace the lid and, if necessary, seal round the edge with self-adhesive tape. Carefully transfer the box, without tipping, to a shelf or cupboard where it can be left undisturbed.

The drying time varies, of course, according to the size and moisture content of the flowers. It can be as little as one day for small single flowers, two days for short sprays such as vetch, and up to four days for hellebores and lilies.

Gently scrape away the desiccant from one flower as soon as you think they will be ready. If they feel crisp and papery, pour the material carefully away. To store the flowers, push each stem into a piece of dry foam (the type you use for flower arrangements) or non-hardening modelling clay (*figure C*). Do not let the flowers rub against each other or they will crush. Keep them in a dry, airy room until you are ready to arrange them. Alternatively, you can pack them carefully into airtight containers.

And how to use these short-stemmed flowers when you have dried them? They can all, of course, be used in miniature work, combined with pressed leaves and other dried material – grasses, hung-dried flowers, cereals and seedheads. Or they can be mounted on to false stems, either pushed into hollow straw (*figure D*) or twisted on to florists' wire (*figure E*). One of the prettiest decoration ideas is to use them in 'close work', the stems pushed into one of the pre-formed foam shapes you can buy, or cut from a block. To illustrate this idea, there is a romantic heart centrepiece shown in *Chapter 6*.

Figure C

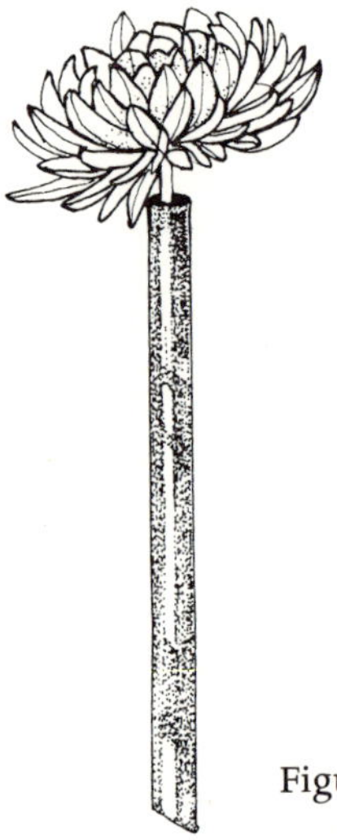

Figure D

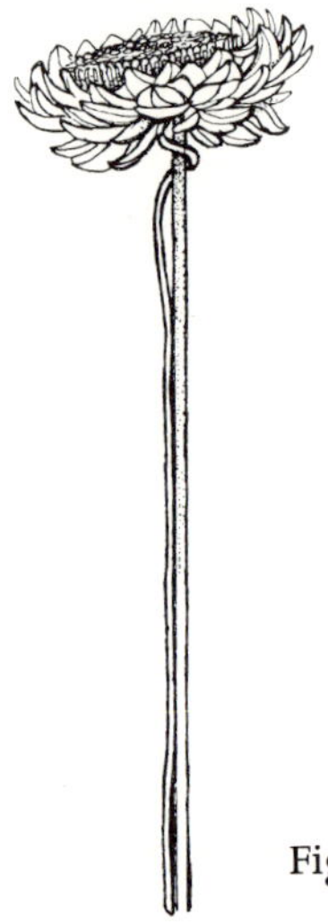

Figure E

Preserving in glycerine

The colour photograph on *page 12* shows the golden, supple lustrousness of leaves and fruits being preserved in glycerine. It is the look of the falling leaves on a glowing, not-quite-winter day, yet the transformation from young greenness has taken place indoors, not on the trees.

The success of this method of preservation depends on the solution being taken up by the stems, right into the leaves; and to do this the stems have to be cut from the trees while the sap is still rising. The time of year will, of course, depend on where you live and the type of climate. Indeed, it can vary a little from year to year, just as the weather itself does.

The formula for the solution is one part of glycerine to two parts of very hot water; or use car anti-freeze liquid instead of glycerine. In this case use the anti-freeze with equal quantities of very hot water. Wherever glycerine solution is called for in this context, you can substitute anti-freeze solution with equally good results.

The plant material takes up the liquid, the water evaporates and the plant cells retain the glycerine. And so it is a nearly natural process, with only a slightly helping hand. No wonder the results look so life-like.

Choose only perfect sprays and leaves to preserve, and cut them when they are just mature. Discard any that are blemished or damaged and any leaves that would come

Opposite, large spikes of foxgloves and lupins, yet still they make a delicate and pretty summer display. Details on page 47
Next page, projected dramatically against a black background, a flower cascade radiating from a trio of buttercups. Details on page 49

To Anne and Michael

below the level of the solution. Grade the material by height. Lightly crush the stem ends with a hammer or wooden rolling-pin or, better still, split them for one or two inches with a sharp knife.

Mix the glycerine or anti-freeze liquid thoroughly with the hot water. Stirring is not enough because of the different properties of the fluids. Shake them vigorously together in a firmly stoppered vessel.

Choose containers in appropriate sizes for the material, 1- or 2-lb preserving jars for small sprays, old earthenware pickling jars for the middle range, up to narrow buckets for the largest. Do not, though, have containers which are unnecessarily wide; they tie up your assets, in the form of the solution, for too long.

Pour in from two to six inches of the solution. Stand the stems in so that they are well immersed – push each one down separately to be sure – and leave them in a cool, dark place. The leaves are fully processed when the undersides become shiny and damp. By then they have taken up as much moisture as they can absorb. Check from time to time to see if the containers need more liquid. If they do, mix more solution, as before.

Remove the sprays as soon as they are ready and wipe off any excess moisture from them. Stand them in empty containers in a cool, dry place. Strain the solution and store it in airtight containers for future use.

The leaves of most deciduous trees can be preserved in this way. Alder, ash, beech, almost more familiar to flower arrangers in their preserved than their fresh condition, eucalyptus (gum tree), hornbeam, Spanish or sweet chestnut and sycamore all give good results, and there are many others. One of the few deciduous trees whose leaves do not preserve well are lime (linden). Mine have always deteriorated into a soggy mess.

Most conifers preserve well, yew and cypress particularly, though some, for no accountable reason, just do not work and the spikes fall off a few days after the sprays are removed from the solution. All ferns preserve well and are wonderfully useful in designs.

One of the great joys of preserving is that you really can suspend a time in the life of the tree or shrub by treating sprays of leaves complete with nuts, fruit or seedpods.

Brambles give exciting results; the berries shrink only slightly and still glisten, like black velvet. The red berry fruits of mountain ash, hips and haws, barberry (bearberry), and the red and yellow holly berries are all naturals for Christmas and other winter decorations. Spray them after preserving with ordinary hair spray which serves the dual purposes of further helping to preserve the berries and restoring some of the original gloss.

Branches of Spanish chestnut with the fruit enclosed in the hard, prickly cover; acorns in their cups; beech mast, sycamore seed 'keys', can all be preserved. Imagine the range of texture you can store up, using not much more than a pint of glycerine and a spare hour or two each weekend.

Very thick leaves – the larger ivies, fig, some ferns, laurel – can be totally immersed in the solution in a shallow dish. They will be ready when the leaves have turned brown. Remove them from the solution, wash them in clear water, gently pat them dry and spread out on blotting paper. To lighten the colour after preserving stand the leaves on a sunny windowsill. Otherwise store them away from direct sunlight.

Previous page, dried sea-shore flowers and leaves make an unusual table decoration, built up on a pre-formed foam shape. Details on page 59
Opposite, in a casual, holiday mood, a brown pot filled with brilliant yellow flags, water grasses and rushes. Details on page 60

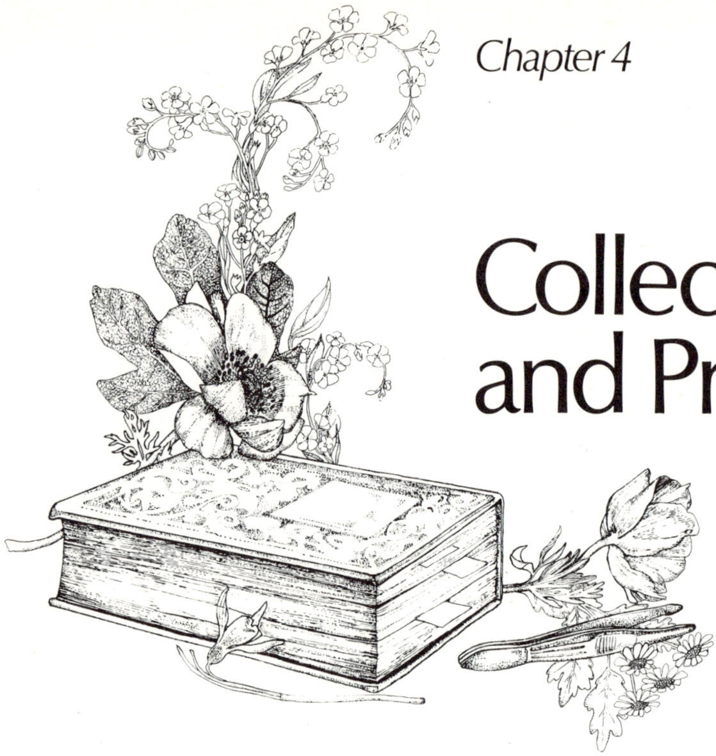

Chapter 4

Collecting and Pressing

AT first sight, the hobby of pressing flowers and leaves seems almost too good to be true. You need hardly any equipment, there are only a handful of rules and the raw materials can all be free. Equally, there are a number of attractive and decorative ways to use pressed materials – not just in pictures and on greeting cards, but even on candles, glass, pottery and china.

You do need patience, though, because flower pressing is not an 'instant' craft. Most materials should be left in the books or presses, in a dry, airy room, for at least four weeks before being disturbed. And it is often a case of the longer, the better, for flowers which have been pressed for a shorter time tend to fade on exposure to the light – blue flowers particularly. But this enforced waiting time is not such a bad thing. In effect it gives one the joy of harvesting in the winter months the collections and memories of spring and summer.

Basically all the equipment you need is a large, heavy book with non-shiny pages, such as an old school atlas or encyclopedia, sheets of blotting paper trimmed to just under the page size, and heavy weights such as bricks. You can buy flower presses, usually with sheets of thick corrugated card alternating with blotting paper, and thumb screws turning on a rod to apply the pressure. These are not essential, though they do have a certain advantage when dealing with thick-centred flowers. The screws can be tightened every day for the first two or three days as the material is flattened and starts to dry, until the whole surface comes into contact with the absorbent paper.

Blotting paper is best for all the middle-weight material, though for very delicate flowers and sprays you can use paper napkins or toilet tissues. Be sure to choose ones which have a completely smooth surface; the crinkled pattern on kitchen-roll paper gives unlooked-for waffle effects. Use newspaper for large sprays of leaves, ferns and grasses and press them under the cushions of a chair or sofa, under the mattress or the carpet.

Material for pressing must be gathered when it is in peak condition, with grasses and leaves the only exceptions. In their case it is permissible to let nature do some of the drying out for you. Cut grasses when they are beginning to turn slightly beige and feel crisp, and wait for the leaves until they are starting to fall from the trees. Not only will the leaves be almost dried, but they will have turned all the rich shades of gold, red and brown that make them such a dramatic and valuable part of a collection.

You can even pick up these leaves on a damp day – in fact, the dampness is an advantage, as it makes them supple and easier to handle without cracking. Wash off any collected grime, wipe the leaves clean and dry with tissue, and press as described. Grasses and leaves picked when they are partly dried will be ready in only about a week.

On the spot

Having dealt with the exceptions, back to the general rule. It is so important to treat material in perfect condition that you must sometimes press flowers on the spot, as you pick them. This means taking a large book with plenty of sheets of blotting paper ready trimmed to size, some strips of paper and a pencil for the labels, scissors to trim the flowers and stalks neatly, and a small paintbrush or pair of tweezers to move the flowers about on the page. If you are picking wild flowers and there are some gaps in your knowledge, take a reference book too. It is so much easier to identify flowers in the round than several weeks later when they are transformed to a single plane. And it is so much more fun to learn about the plants you are handling and preserving.

Ideally, flowers for pressing should be gathered in the middle of a dry day, when the morning dew has had time to dry off, and before the evening mists come. But this is a counsel of perfection and poses all kinds of problems for the majority of people. These conditions, when you will find flowers with no excess moisture to be dried off, give you the best chance of good results. But flowers cut at the only available opportunity, be it just after a storm or late in the evening, are preferable to no flowers at all, and you might be lucky. In any case you have little to lose by experimenting with them.

It is best to keep one type of flower, or those of much the same thickness, on each sheet. Flowers and leaves vary considerably in moisture content, and this governs the drying time. Added to that, very sturdy flowers – those with thick centres – make a thick sandwich between the sheets of blotting paper. The curves that they take up would prevent the absorbent paper from enclosing finer flowers on both sides. Set the material out on the page neatly, making the best use of the space yet not allowing them to touch each other.

Generally, and not surprisingly, it is the delicate looking flowers which wilt most quickly and therefore need pressing *in situ*, unless you cull them from your own garden. Violets are an obvious example. They wilt practically as soon as you look at them. Forget-me-nots, harebells and stitchwort are others. Even fairly large flowers such as the vibrant ragged robin have a short life unless they can be put straight into water, and so they might well come into the on-the-spot pressing category. It is often easier to press flowers in a friend's garden or a field than to carry them home in a slopping container of water.

Place single violet flowers face down on the page for pressing. Encourage a gentle curve into the stem when pressing sprays, and cheat a little by folding flowers over when you want the effect of a half-flower or bud. This gives greater variety when you come to use the flowers for designs. Forget-me-nots are best pressed in sprays. If you want single minute flowers in your design you can snip them off afterwards. The open, cup shape of the white stitchwort and the long trumpet shape of the white rock-cress both need care at every stage. Push them around or the page with the bristle end of a paintbrush; even the gentlest handling with tweezers could prove too much for them. And, remember the saying 'a puff of wind will blow them away'? It is never more true than of a page of tiny flowers. On a windy day if you are collecting in the country, sit in the car or a friendly café; try to rig up some kind of windbreak, or at the very least sit so that you protect the book from the direction of the wind.

Much easier to handle, and possibly better subjects for complete beginners and/or children, are all the flowers with flat surfaces – daisies, buttercups, the potentilla (cinquefoil) family, wild strawberry and so on. These will be fairly happy to be taken home for pressing, although it does take only a few minutes per page to press them where they grow. For best results, place the single flowers face down on the page; this helps to flatten the slight inward curve of the petals and makes it easier for you to press a gentle helping thumb on to any thickish centres.

In a different category come the 'bunch' flowers like the clovers, which also press well. Because of their round shape, they need more pressure and might take more than the minimum four weeks to press. White clover turns only slightly parchment coloured and can produce pressed flowers with two totally different shapes. One is a complete circle with two green rings in the centre, and the other a much fatter semicircle, barely recognizable as having the same origin. Pink clovers are just as successful technically but perhaps leave something to be desired aesthetically. They tend to turn a not very attractive shade of brown, as if they had died on you. If this offends your eye, stick to the white ones.

Clover stalks are one of the most useful types of all, so it is always worth pressing the flowers complete with stalk. Then, if you want to, you can use the two parts individually after pressing. Or press the flowers and stalks separately, again helping the stalks into barely discernible curves. For very few plants grow ramrod straight in real life and designs using dead straight lines tend to look forced and unnatural.

Other useful stalks to press are buttercup, larkspur, traveller's joy (clematis) and all the grasses. As you build up a collection of flower sprays, you will find yourself left with lots of small pieces of stalk which all, at one time or another, will have their uses.

All the umbrella-shaped flowers, such as cow parsley and chervil, are excellent subjects for pressing. They look like creamy-white snowflakes, or hoops of minute white stars. These flowers can be snipped off into tiny separate fragments, or used whole, perhaps with another flower in thc centre, covering the wheel-like stem formation. They look especially stunning against a dark background; the deep midnight purple candle in our colour photograph on *page 66*, for instance.

It is worth taking time and trouble to press several pages of these flowers. Indeed, the cow parsley (wild chervil) plant is one of the richest sources of material for pressing. The leaves, which look rather ordinary and unassuming on the plant, turn purple mottled with bright green when pressed, and hold this colour combination seemingly indefinitely. Again, the leaves can be used whole or snipped into tiny leaflets.

Small sprigs of elder flower, too, give a snowflake effect. When pressed they turn a deep rich cream colour and are fluffier than chervil.

Never be put off by the size of a flower. Some of the largest ones in my collection are the most impressive and look like shapes of pure, fine silk. Ordinary pink and white bindweed, for example, take on a look of pure luxury when pressed, emerging like glowing trumpets, the colour ranging from the deep brown of the sepals, through copper bronze to almost milky white at the top.

Poppies, too, can offer some pleasant surprises. The traditional way to press them is petal by petal, reassembling them to make a whole flower on the design. But pressed whole, complete with black and yellow seeds and golden stamens, they are unequalled. Depending on type, the petals turn through a range of deep reds and purples which, if not actually an improvement on the original vibrant shades, are at least attractive colours in their own right. One perfect poppy on a white candle makes this point. *See the colour photograph on page 66.*

You have to use a certain amount of judgment when pressing large flowers, for clearly their own inclination, on the paper, is to crease up like soft folds of cloth and not to emerge from the process looking like flowers at all. They unquestionably need a little more care and attention than others, so it is a good idea to press them in the pages of a small thick book, one flower to a page.

Take poppies. To preserve the round shape, push the flower gently but firmly in the centre, bending back the bell-shaped calyx until it is in line with the stem. Then arrange the petals round, overlapping where they naturally fall and taking care not to snap them off at the base.

Trumpet-shaped flowers like convolvulus and bugloss cannot be persuaded into quite this arrangement. Lay them carefully on the page with the underneath petals flat against the paper and press the topmost petals gently down over them. This gives, in pressing, a very realistic view of the flower; just as it appears from the side when it is growing.

Bell-shaped flowers on a small scale, like harebells and hyacinth, can be pressed in this way. Or they can be sliced in half lengthways, with a very sharp knife, and each half can be pressed separately. This is also the way to treat daffodils, and you get two pressed flowers where you had only one.

Honeysuckle flowers are a much more awkward shape and the pressed flowers can look rather ragged. The colour is so lovely that it is worth dismantling some and pressing the petals separately. You can always do this, of course, with any flowers. The honeysuckle petals spread a little, but retain the characteristic inward curve and are easy to reassemble into the original flower structure.

Leaves have a vital part to play in designs and must be preserved with just as much care as flowers. Indeed, many designers tend to use leaves and flowers in the proportion of two to one. To build up a versatile and varied collection you have to train yourself to look beyond the flowers that so often steal the limelight and take a critical view of the leaves.

Freshness is crucial with all but the falling deciduous leaves, though most of them usually offer you at least the timespan of the journey home before they need attention.

You can look to leaves for four basic types of construction: long points, such as willow and lungwort; soft circular shapes, from the trefoil of clovers to the cascade of larkspur; ovals, like ash and periwinkle, and what might be called ivy-leaf shapes, which would include maple.

You will find almost as much colour variety in a collection of leaves as in flowers. But, although a writer can give general guidelines about the colour range to expect after pressing, it is by no means predictable with certainty. It is this variety which makes the hobby so fascinating, and the time when you can open up the pages of your pressing books such an exciting one.

Lungwort leaves are hairy on both surfaces and so, when pressed, have a slightly speckled look. The shape is very much that of willow, though of course they are much less substantial. More pointed still, stitchwort leaves retain a sharp acid greenness which is seen to good effect against cream or pale grey backgrounds.

By far the most spectacular of the round leaf shapes is larkspur, like spinning fireworks or giant cartwheels. Choose the well-developed leaves which are close and thick, and they will make patterns on your designs like motifs of the finest lace. Sprays of the very smallest side shoots, complete with flower buds, are quite different. The little clusters of leaves press against the stem like outstretched hands and look soft and feathery. Some of the dead-nettles and wild mints have round leaves of more solid characteristics, and there are many more.

The oval leaves cover a vast range of sizes. There are members of the pea family, the vetches, with small oval leaves in pairs along a curving stem. Others as far apart as rose and great willow herb give the same general outline in pressed flower work.

In the ivy-leaf shape class, currant leaves dry to a deep holly green on the face and a paler greyish-green on the reverse. But what is most striking about them is the deeply etched vein pattern, invaluable in designs when you really want a leaf to look like a leaf!

Ideally maple leaves should be pressed when they are very young and small, in pairs or in clusters from the tip of the stem, with some of the leaves as yet undeveloped. They dry to a range of deep mahogany brown colours, with some reds and dark greens among them; even providing some complete surprises when the same shrub branches out into a variety of pale cream and light green colours.

The photograph on *page 21* shows a page of pressed flowers in a child's book – a late nineteenth-century edition of *Girl's Own Paper*. The flowers, from the top, are white clover, then some of the flowers on stalks, alternating with daisies; a row of honeysuckle, then sprays of bellflower. The leaves, from the top, are herb Robert, maple and vetch. The page of tiny flower sprays is speedwell. Labels identify each type of plant material by the Latin names.

Identification of some kind is important when you remove the pressed materials from the books or presses. Depending on the extent of your interest, you can label the flowers with their common name and Latin names, or categorize them according to size and colour, simply marking the envelopes 'large red flowers', 'small blue sprays' and so on.

Even when natural materials have been pressed they are capable of reabsorbing moisture; and this means developing mould. It is important, therefore, to store them in a dry, airy room or in airtight containers. You can use ordinary envelopes or transparent ones, one to each flower and leaf type, or store the materials between layers of tissue in a cardboard filing box, unused drawer or old dress box.

Design time

When you want to make a pressed flower picture it is a great help to have the frame first. This might sound like putting the cart before the horse but it has a two-fold advantage. First, having the frame, be it round, oval, square or rectangular, imposes certain limitations on your design. And, particularly for a beginner, this is a great help. With the whole world at your command and nothing decided in advance, it is rather difficult to know where to begin. Secondly, it is useful to have the glass to cover your design at all stages. When you have to leave it for a few moments or a few days it is comforting to know that your work is protected. If you have old frames and they are missing glass, get a piece cut to fit before beginning the design.

You must never put a raised mount round a pressed flower picture; it is imperative for the glass to come into direct contact with the natural materials. It is this close contact which keeps the delicate petals and leaves in good condition. Even when you make greeting cards, you should cover the surface with transparent film so that the flowers stay well preserved.

The best background to use when making pictures is artists' mounting board. It is available in a wide range of strong and subtle colours and there will surely be one which will exactly suit the feeling you are trying to create. Have the board professionally and accurately cut to the inside measurement of the frame. Or perhaps you can do this yourself, using a steel rule, set square and craft knife. Have a piece of hardboard cut the same size, to use as a backing. Then when the picture is finished you can achieve a really tight sandwich of glass, pressed flowers on mounting board, and hardboard.

Use an adhesive with a latex or rubber base – Cow gum or rubber cement. You will need only the tiniest scrap at the centre of each petal or leaf, or at intervals down the length of a stem, for remember that the design will be constantly under pressure from the glass and this will keep the materials in place.

There are designs using pressed flowers in *Chapters 5, 7* and *10*.

Chapter 5

Country Gardens

IT was as long ago as 1213 that Alexander Neckham, a horticulturist who graduated in Paris and later became Bishop of Cirencester, gave his recipe for the perfect garden.

> 'The garden should be adorned,' he wrote in *De Naturis Rerum*, 'with roses and lilies, turnsole, violets and mandrake; there you should have parsley and cost, and fennel, and southernwood and coriander, sage, savory, hyssop, mint, rue, dittany, smallage, pellitory, lettuce, garden cress, peonies. There should also be planted beds with onions, leeks, garlick, pumpkins, and shalots; cucumber, poppy, daffodil and acanthus ought to be in a good garden. There should also be pottage herbs, such as beets, herb mercury, orach, sorrel and mallows.'

And it was a sensible plan for large gardens, a carefully balanced mixture of vegetables, medicinal and scented herbs and flowers. People who tended small plots had to work out a careful balance, too, but with an even greater emphasis on food, for these gardens had to provide the staple diet for the family.

There must have been a considerable overlap in the plants they grew, for cottagers had only two sources of acquiring them: they could wait to be given seeds and cuttings from the monasteries and large estates, or they could introduce the plants growing wild in the fields, woods and hedgerows all around them.

And so the difference between the large and small garden lay not so much in what they grew as in the way they grew it. In large gardens, space could be used to create different visual effects, according to the fashion of the day, with sweeping lawns, mazes, sculpture, secret walks, pools, ornamental lakes and extravagant, frivolous buildings.

Opposite, the vibrant colours of a print of the Mississippi are repeated in the flower and shell decoration. Details on page 67

MIDNIGHT RACE ON THE MISSISSIPPI.

WITH

Space was at a premium in small country gardens. It did not matter what they looked like, as long as they were fully productive. Onions, leeks and shallots there would be; cabbages, beans and, later, potatoes. But not in beds. They would be jostling for position with all the herbs, and with salads such as violets, primroses and marigolds. Many of the colourful flowers were an incidental by-product, as it were, of a crop grown for food. Where other flowers were introduced, such as foxgloves, cranesbill, campions (catchfly) and roses, they would have had to take their place in an already tangled web.

Oddly, it is this very characteristic, the total lack of any concession to planning, except for the table, that gave the small country garden the charm that has persisted through the ages. With plants growing as close as could be to their wild state, and to each other, a virtue was, in a very real sense, made of necessity.

One of the most delightful features of these old-fashioned gardens must have been the scent. Imagine the pleasure of wandering out into the dusk on a warm summer's evening; savouring the richly varied bouquet of all those flowers growing together, evocative and exotic as an Eastern spice trader's cargo.

In the great houses of Europe in the sixteenth and seventeenth centuries, one of the most romantic positions was that of The Lady of the Pot-Pourri, whose enviable task it was to see that the rooms, clothes presses and linen were always deliciously scented. And for the raw materials of her art, of course, she drew on the kitchen and herb gardens and the flowery parterres and knot gardens.

The mixture she prepared would have consisted largely of rose petals – one of the most lingering fragrances of all – and pungent leaves, with the addition of various oils, spices and fixatives. You can still make pot-pourri in this way, though the ingredients are becoming increasingly difficult to obtain.

Alternatively, you can recapture and hold those scents of spring and summer by making a mixture of colourful flowers and petals, as decorative as a Victorian posy, and 'fixing' it with a modern chemical preparation. This powder (Bio Pot-Pourri Maker) contains oils of lilac, violet, lily of the valley, orange blossom, rose, lavender and rosemary; spices like cinnamon which are readily obtainable and others, like frankincense and patchouli which are not; and the fixatives, oak-moss and vetiver root.

When making pot-pourri it is not essential to choose every flower and leaf for its scent alone; the oils or the powder you add to the mixture will give additional fragrance. There is room, therefore, for some flowers or florets just because they are a beautiful shape or a pretty colour.

Collect flowers, petals and leaves as each type comes into blossom or reaches peak condition. Choose larkspur florets and cornflower petals for the lovely deep blue; hyacinth florets for the range of colours, from palest pink to deep purple; tiny wild or cultivated pinks, from the very softest to the very darkest of their colours; the little pea-like flowers of lupins and, of course, rose petals galore. Gather any scented leaves you can find, such as thyme, marjoram and all the mints. Strip the florets from the stems and gently pull away the petals of larger flowers; tie the herbs in bunches and hang them to dry. Spread out the flowers on sieves, wire netting

Opposite, yellow lupins naturally twist and turn. Here they are arranged in an old iron charcoal burner. Details on page 67

or newspaper and leave them in a warm, dry place, out of direct sunlight for a week or two. Stir them over with your hands occasionally. When the material is dry it will feel as crisp and crackly as paper. Carefully strip the leaves from the stems, being sure not to crumple them into a powder. Store the dried flowers and leaves in a lidded container or a tightly sealed polythene bag, or in separate containers according to the colours if you want to make mixtures in individual colour ranges. Add more flowers and leaves as they become available.

When you have a mixture which is varied and colourful enough – and, after all, the term pot-pourri means 'medley' – you can add the other ingredients. If you choose to use the commercial preparation, simply follow the manufacturer's directions. That is to say, put the dried materials into a large container, such as a polythene bag, add the powder and shake thoroughly. Leave the mixture to mature for at least two weeks.

If you are lucky enough to live near a specialist shop which stocks the traditional pot-pourri ingredients, or you can obtain them by mail from a herbalist, here is a recipe close to one which is four hundred years old. I am bound to say that I find this traditional method more satisfying and therapeutic to use than the other.

Dry about 5 handfuls of rose petals and put them in a lidded container with about 4 handfuls of salt. Stir them thoroughly at least twice a day for 5 days – this is one of the great pleasures in the whole process. Meanwhile, gather and dry an equal quantity of other flowers, petals and leaves (see the suggestions above). When they are dry, add them to the rose petals and blend well together.

To this mixture add these ingredients: 1 oz of flower oil – oil of lavender, rosemary, geranium or whatever you can obtain, or a mixture of these; 2 sticks of cinnamon, crushed; 1 oz ground cloves; 1 oz ground nutmeg; 1 oz ground coriander and 4 oz orris root powder, the fixative.

Put the orris root powder into a cup and gradually add the flower oil, stirring well. Add this paste and all the other ingredients to the flower mixture and stir thoroughly. Leave in a covered container to mature for about four weeks, stirring occasionally.

Now that you have a long-lasting flower mixture, put it into the most decorative containers you can find. One suggestion is shown in the photograph on *page 111*, an old stone mortar decorated with pressed poppy petals, cow parsley (chervil) flowers and daisies. Open bowls of any kind are suitable, Chinese porcelain ones especially, though naturally the fragrance does not last as long in uncovered containers as it does in lidded ones. Consider filling lidded glass jars, scent bottles, ginger jars or an old teapot, and making sachets. Fine cotton envelope sachets or tiny draw-string bags are pretty. Lift the lid from one of the containers and it is like opening a window on to the kind of country garden the cottagers knew. Open a drawer where one or two sachets of pot-pourri are nestling among the linen, and you can make believe that you employ your own Lady of the Pot-Pourri.

Kitchen herb jug *see colour photograph on page 22*

It seems only fair to let you into a secret: the arrangement of herbs in an old green and white jug started by being simply a handful of fresh herbs brought in from the garden. But somehow the deep green against the pine and brown of the kitchen, the interplay of spiky and round shapes, soft and firm textures, all looked too good to eat – at least before they had been photographed.

For the arrangement, the jug is filled with pre-soaked foam holding material. The fan shape of the design is outlined first with long stems of sweet-smelling lavender, and the height determined by a thick, bushy stem of rosemary. Two shorter stems of rosemary, within the curve made by the lavender, give weight at the sides. Long, trailing stems of ivy leaves spill way out of the jug and over the working surface, forming a screen through which the pattern on the pottery is just visible. There is also, for good measure, marjoram, thyme, mint, herb Robert and yellow chamomile complete with its daisy-like flowers.

Herbaceous arrangement *see colour photograph on page 31*

At first glance, the display looks as elegant and formal as one you might find gracing a wedding buffet or charity ball. Indeed, it would not look out of place at either event, though it might surprise some of the guests to know that the flowers had been gathered almost entirely from a wild stretch of beach and a patch of waste ground.

The flowers fall into three colour groups – mauve, pink and cream – and three shapes, long, flat and round. Within these limits, it is easy to create a display that has movement and interest from top to toe.

The container is a shallow white pottery urn, a fairly classic shape, but any pedestal dish, such as an old cake stand, would do equally well. The urn is filled with pre-soaked foam holding material (Oasis) and covered with crumpled chicken wire mesh tied securely with string. This precaution is essential when an arrangement is going to consist of long, heavy material. It is also advisable to cut all the stems slantwise, at a sharp angle, to help them push smoothly through the foam.

The arrangement is basically a triangle shape, with the central line and height provided by foxgloves. The tallest stem draws a line of deep pink, becoming paler towards the topmost buds, right through the centre. The points and steep sides of the triangle are defined by shorter stems of foxglove and spears of woundwort (betony), and the two lower points by downward-curving stems of tree lupin, shaded from milky white to deep cream.

In the centre, pride of place is given to a large, round head of one of the umbellifers, which is ringed by pinks and the deeply coloured wild pea sprays. Huge, hanging heads of elder flower, like misty clouds, provide a background to silhouette those flowers with more positive outline and shape, notably the deep mauve of meadow cranesbill. The white and pink flowers cascading over the sides of the urn are campion.

Pink and pewter

You can imagine the flowers in the group shown in the black and white photograph overleaf in any of your favourite colour combinations – it could be in shades of white and cream, orange and yellow, blue and mauve, but the original was, in fact, in pinks, from the very palest sugar-almond shade to the deepest nearly magenta tone.

There is just as much variety in the shapes of the flowers chosen, too. There are the long, slender and elegant trumpet shapes of freesia; the flat, simple daisy formation of

Michaelmas daisy; tight, hard sprays of double stock; velvety cones of antirrhinum; drooping lanterns of wild fuchsia; clumps of wild marjoram, and the flat, disc-like heads of an ice plant.

All the flowers in the pewter mug were picked from the same border in a country garden at the same time – some were flowering late in the season and others were putting in an early appearance. With such a wealth of colour and shape, the arrangement seeks to do no more than present the flower medley so that each one is seen clearly and well. This means alternating pale and dark colours, well-defined shapes and clustered masses all the way round.

The holding material is a block of soaked foam extending well above the rim of the container. The outline of the shape was made with a ring of the flowering wild marjoram, since it was the most plentiful material to hand. Then the freesias were arranged, some at the back, some at each side and a spray of the buds extending through the centre.

A large disc of ice plant begins the centre focal point, encircled by small sprays of the daisies and long stems of antirrhinum. Stocks, light and dark, single and double were placed next where they do most to contrast with the depth of the marjoram, and lastly the sprays of wild fuchsia were placed to hang low at the base of the design and form a veil over some of the darker colours.

In the foreground of the photograph there is a small arrangement in the same colours, but here all the flowers are dried. The container is a crystal inkwell with a brass-rimmed lid. No longer serving a practical purpose in the household, but a welcome light relief on a busy office desk. A mound of Plasticine (non-hardening modelling clay) is pushed into the aperture and shaped to form a dome above. A diagonal line of helichrysum (straw daisy) flowers defines the shape from the top left to the lower right of the design and is emphasized with two heads of hare's tail grass, dyed deep cyclamen pink. The outline is extended with sprays of dried rue seedheads, now a deep ginger brown; trailing over the front of the inkwell, there are a few sprays of lino grass, dyed black. Single pale pink xeranthemum flowers nestle beside the darker pink of the helichrysum, and pale brownish-pink florets of hydrangea (dried by standing in a container) cover the holding material all round. For short, sharp emphasis there are a couple of pink-dyed barley heads and for substance two small poppy seedheads.

On the right of the photograph, a deep purple candle gives a preview of designs to come – in *Chapter 7*. It is decorated with pressed flowers and leaves: a circle of larkspur leaves, a complete pink in the centre, a whole musk flower and a scattering of separate petals. Pressed poppy flowers and more larkspur leaves on the table are ready to make more decorations. Full instructions are on *pages 79 to 82*.

Flower cascade *see colour photograph on page 32*

So many flowers of the romantic, nostalgic, country garden type are ideal subjects for pressing that any collection is sure to provide an overwhelming choice. Indeed, the problem is likely to be what to leave out.

The picture in the photograph is a crescent shape worked entirely in mauve, yellow and white, the traditional colours of spring. It is a study in apparent casualness but, as you might imagine, to achieve this random look there has to be a firm basis of careful planning.

All pressed flower designs – all floral art, in fact – must have a focal point, the place where the eye is first drawn; otherwise the composition will lack coherence. In this case it is the cluster of three buttercup heads which form both the brightest colour point and the strongest concentration of any single colour. The stems giving the upward and outward curves to the design all appear to radiate from this spot.

Before starting a piece of work of this kind it is helpful to sketch out the general outlines, drawing, if you like, broad sweeps around large dinner plates, smaller curves around saucers and the positions of single flower heads around coins. It will be easy to see at a glance from this simple presentation whether the balance looks natural and pleasing, or whether the design is inclined to be top heavy or too cramped, or leaves vacant unreasonably large areas of the background. Do not be afraid to use your judgment at this stage. If it looks wrong to you, then it is wrong *for* you. When you have an outline drawing you like cut a piece of paper or card in roughly the same colour tone as the background and copy the sketch on to it. Place the leaves and flowers in position and move them around until you have the effect you want. It is much easier to have second and third thoughts at these initial stages than later, when there is glue on the flowers and the background.

The background to our picture is black artists' mounting board, 14 × 19 in (36 × 48 cm), the strongest possible contrast to the pale colours of the flowers; and the frame a simple sliver of silver-painted wood, a perfect complement to the greyish green leaves.

The design, then, begins and ends with the cluster of buttercups. This group is surrounded by mugwort leaves, used upside down to show the silvery-grey undersides. Sweeping up to the right, the main height is sketched in by an inward curving spray of common agrimony (Joe Pye weed), the colour coming here and there, with some buds which had not opened and some flowers which had just finished. Looking closely at this stem, it is hard to believe that the flowers are pressed at all, they look so natural.

The agrimony shows the way for the other stems to point, and the line is followed on the right by a frondy piece of meadow grass and on the left first by the white greater stitchwort and then the mauve dead-nettle. Notice that some of the nettle leaves have been pressed flat, to emphasize the heart shapes, and others folded over in half. On the left of the nettle there is another wisp of grass, a single daisy, and a snippet of mugwort.

The mauve flower, so perfectly picking up the colour of the nettle, is a geranium, with stamens of buttercup in the centre. Above that, a leafy sprig of speedwell, sprays of moorland crowfoot and small snippings of mugwort.

Working outward from the buttercups again, the main outline on the left of the picture is defined by a single white clover on its stalk, a stem of meadow grass, mugwort leaves and a spray of wild strawberry with one complete flower and several buds. The scattered flowers, applied after the stalks and sprays have been placed, are clusters of elder flower, single speedwell flowers and greater stitchwort. Sprays of this white flower curve downwards from the buttercups. On the right they are given 'false' leaves of geranium and beyond them there is a sprig of speedwell.

With the basic shape completed, other flowers are added to soften the outline. These include buttercups, daisies, white campion, deepened to cream after pressing, and bellflower.

Back in fashion *see photograph opposite*

It was a charming Victorian custom to send greeting cards made up of pressed flower designs, and it is a custom that has just as much relevance and appeal today. It would be a thoughtful gesture to send such a card to a friend who had given you the pick of her garden for flowers to press, or to other friends without gardens of their own.

Pressed flower cards are so pretty that they are usually for keeps – it would be a heartless person, indeed, who consigned one to the waste-paper basket. And so it is necessary to protect the flowers from damage, dust and moisture. You can do this either by covering the face of the cards with transparent film – there are a number of brands on the market – or putting them behind glass in a frame. We decided to design our cards not to go through the post but to be a permanent decoration, and so we made them specially for framing. One is a Christmas card and the other brings New Year greetings – so what could make a better Christmas and New Year present combined?

The frame is a new maple one with a golden 'slip' frame. You can find moulds like

this in most frame-makers' and have it copied, or you might find an old frame which simply needs a new piece of glass cut to size.

The backing in the frame is covered with a piece of pink dress-weight cotton fabric, which seemed suitably pretty and romantic for a pressed flower notion. Whenever you are covering a backing with fabric, be it a piece of stiff cardboard for a greeting card or a piece of hardboard in a frame, it is vital to achieve a completely smooth surface; even the slightest wrinkles will be all too obtrusive and give a couldn't-care-less appearance to your design. The best way to achieve this tautness is to stitch the turned-over excess fabric across the back of the cardboard. This gives a much better finish than sticking. Cut the fabric at least 1 in wider all round than a piece of card, or 2 in for a full-sized picture. Place the fabric right side down on a table and centre the card or hardboard on top of it. Turn the excess fabric over, neatly fold under each corner and secure the mitred edge with a few stitches. Then, with strong thread, sew the overlapped fabric on the reverse, up and down and from side to side, with large stitches until eventually the backing is firmly and tautly held in. Pull the thread as tight as possible as you go. You do not have to be a neat or patient needlewoman to do this. The stitches can be as large as you like and they will not show.

Now that you have successfully covered the background, you can work on the card designs. Sketch out each one in rough on a piece of paper the same size as the card mounts. In our original designs, each card measured $5\frac{1}{2} \times 8$ in (14×20 cm), with cut-off corners.

Naturally, because of the comparatively small scale of greeting cards, you will be using small and sometimes delicate pressed flowers and so it is even more important to put only the faintest smear of adhesive on the petals. With very small flowers, it is even better to apply the adhesive to the card. If by any chance you make a mistake and have to move a flower or leaf all is not lost. Once dry, rubber-based adhesives can be rolled away with a finger-tip and will not leave a trace.

Christmas card Carefully following the outlines shown in the photograph, stick down the curves of pale blue flowers. Our design shows forget-me-nots, pressed in small sprays and now reassembled to make larger ones, with bellflowers on the left and speedwell on the right. The focal point is the group of two small, well-matched maple leaves and three silverweed flowers, not quite such a strong yellow as buttercups. Two maple leaves of different sizes give weight to the top left, with one of them making a strong background for a snipping of forget-me-not. Two large mauve flowers echo the colours of the sprays, one a wild geranium and the other a campanula with a daisy stuck in the centre. The leaves are snippets of herb Robert and a minute oak leaf.

New Year card This design has a similar shape, an open oval. Sketch the shape before starting to work.

First place the two bramble leaves at the top left-hand corner: the focal point in this card is at the top of the design. Then position the lesser bindweed flowers and buds, encouraging them to follow the curves. Next, in the centre, comes the geranium flower, then a single buttercup head. Buttercup stalks curve upwards from this, one given a scattering of golden vetch flowers and the others two yellow violets. These flowers retain their colour well and are good value in a collection. The two leaves are snipped from brambles. The upper curve is composed of a spray of bellflower, with 'false' leaves, a spray of vetch leaves and speedwell flowers.

To obtain a professional look with the lettering, buy a sheet of rub-off instant lettering in a style you like. Most good artists' materials suppliers stock them. Count out the letters you need and the space you have and take care to get the letters and words evenly spaced on a piece of thin card.

Glue the flower designs and message strips on to the cotton background and edge the message with four russet leaves. Put the design in a frame and seal all round the edges at the back with sticky brown paper strip.

Opposite, in a different mood, lupins are persuaded to twine in and out of a piece of bleached driftwood. Details on page 67
Next page, the simplest possible presentation for colourful, exotic blooms of vibrant pink rhododendron. Details on page 16

Chapter 6

A Hint of the Sea

BEACHCOMBING. Treasure hunting. Looking for rocks and stones and shells and driftwood. And flowers. All the soft blues, pinks, mauves, yellows and whites, so many of them pastel colours that seem to be shrouded in early morning mist, or a light coating of sea salt. Flowers that are washed by the spray, and either baked by the sun or lashed by the biting wind. These conditions tend to make sea-shore flowers hardy; tougher versions of those that grow inland. And so they are, in the main, good subjects for drying, with little tendency either to shrivel or fade. Most of them last well in water, too, and are therefore worth taking time to arrange.

Even if you do not live within a day's drive of the sea, you can grow or buy flowers and grasses similar to those mentioned here, and copy the mood and feeling of the arrangements that illustrate this chapter. After all, it is not any particular species you are seeking, only designs with a hint of the sea.

There are some dramatic blue flowers, with perhaps viper's bugloss outstanding, its trumpet-shaped flowers the colour sky and sea always should be, if there is justice, in the summertime. And there is a coastal member of the borage family, northern shorewort, sometimes known as the oyster plant, with great clusters of blue and pink flowers hanging down like lanterns. Sea holly has a habit of being a bit off-putting because of its vicious spikes, but there is no need to be beaten by them. The plant dries just as it is, losing none of its subtle colouring, and is invaluable in all kinds of decorations. Being of a chunky nature, it fills up round masses of space particularly well.

Previous page, a picture in deep dimension composed of nuts, cones, acorn cups, seeds and seedpods in symmetry. Details on page 70
Opposite, hanging in a copper ladle against the golden light, faded beech leaves in natural silhouette. Details on page 74

Pink is a familiar colour for sea-side flowers; you might discover sea pea or wild pea, this last sometimes to be found on wet sea cliffs. Or long, fleshy stems of sea rocket and sea stock, such a delicate combination of pale, silver-green stems and leaves and straggly four-petalled flowers.

It might be a little surprising to come across bindweed making its way purposefully across some beaches, but there is a sea-shore variety of this, too. Sea bindweed is one of the best looking members of the family, with huge flowers, like pink and white striped candy cones, which press like a dream. Sea pinks (or indeed any pinks) are among the most valuable additions you can make to a collection of dried material. They lose a little of their colour but gain something almost mystical, the perfect domes like crystal paperweights with the amethyst pink flowers buried deep inside. You can use the flowers on the full length of the stems, or snip them off near the heads for flat work (see the heart-shaped arrangement which follows).

Think of the sea-shore and you inevitably think of mauve flowers, with sea lavender surely the first to come to mind. These flowers dry perfectly by hanging (*see them in the photograph on page 10*) and retain all their original characteristics. They are like close clusters of tiny cones, in the softest of colours, and last for ever. You can buy sea lavender – or statice, as it is also known – in many florists' and department stores, even in gift shops, often in a number of pale colours. It is a very worthwhile investment to indulge yourself in a bunch of each colour there is as it can form the very foundation on which so many designs are built.

There are several yellow 'creepers' that inhabit the sea-shore. In some places you might find yellow vetch, with sprays of the pale yellow flowers that dry so well by the crystal or powder method; wild cinquefoils or potentillas with silver-green, fern-like leaves, and bird's-foot trefoil, with clusters of bright yellow-orange flowers.

If you are lucky enough to find a stretch of coastline (or indeed any sandy in-shore area) where the tree lupin has established itself, you will be able to pick cream-to-yellow flowers for drying – they are one of the most delightfully fragrant ingredients of a pot-pourri – and for fresh arrangements. The flowers are capable of lasting in water for well over two weeks. Collect the seeds if you have sandy soil; they are not at all difficult to grow.

Look on some sea cliffs and pebbly beaches for sea kale, whose white four-petalled flowers contrast so sharply with the dull greenish-grey of the cabbagey leaves. The flowers are surprisingly attractive. Sea campion bears single white flowers which are pretty in water or can be pressed to a soft milky-creamness. Mignonette has minute white flowers closely packed round tall stems, like huge shining spikes; and marsh mallow, which gives its name to the confectionery, has showy, snowy-white flowers nestling amongst leaves folded protectively around them.

The grasses, rushes and sedges which abound on the coastlines of every continent are a study in themselves. And, curiously, none of the flowers would seem as beautiful without the unassuming backdrop these provide. Follow the directions in *Chapter 3* for drying them.

Two of the most decorative of grasses, worthy of special mention, are hare's tail, with large soft cotton-woolly heads practically the colour of the sand itself, and large quaking grass with red-tinged heads, something like hop flowers, hanging so heavy that it is a wonder the stems hold them. Gather as wide a variety of grasses as you can, some with statuesque formation to give both height and solidity to your arrangements and some softer, lighter ones for smaller designs.

There are some dramatic, rich browns among the rushes. Look for ones with dark chocolate brown heads, cascades of brown flowers like a sepia print of a fireworks display, and some with soft brown downy flowers, like the babiest of birds' feathers. Bear in mind that you do not have to use even the longest of rushes in grand or vast groupings; you can snip off a piece any length you like.

Flower heart *see colour photograph on page 33*

Generous bunches of sea pinks and sea holly hanging up to dry might bring to mind a leisurely stroll along the beach, a day when you filled up your baskets with a salty, sandy harvest. And so it might seem a little out of character to find these two materials the centres of attraction in a romantic heart-shaped flower decoration. This would make a lasting gift for a couple celebrating their engagement or wedding; or a charming memento of a family silver wedding. All the flowers are dried and therefore virtually everlasting and so the decoration can be a table centrepiece now, a door or wallhanging for Christmas or Thanksgiving Day later.

You could cut the heart shape from a block of foam, but we decided to have this one made up of wired moss by a florist. It is almost 18 in (48 cm) long, but of course, being custom made, could have been any size required.

For a decoration of this type, choose a selection of dried material within a limited colour range – ours is blue, brown and pink, not too romantic or feminine – and in a rich variety of contrasting textures. Especially when the colours are muted and somewhat understated, you need the contrast of shiny and matt, smooth and spiky material.

Figure F

First, the shape is outlined with a tightly packed row of pine cones, each one twisted on to a short stem of florists' wire (*figure F*) and pushed well into the moss. Cones are heavy and, particularly if the decoration is to be used as a hanging, need securing firmly; otherwise their own weight will literally be their downfall.

Next comes a clumpy row of prickly sea holly, almost the colour of pale sand in the moonlight. You might need to use gloves to handle it, or at the very least a long pair of strong tweezers or pincers. Next, deep mauve thistly knapweed flowers and pale silvery sea pinks.

By contrast, the next row is ginger brown, the upturned cases of beech mast. This colour is given emphasis by a double row of matt black hogweed seedheads (any large umbellifers would do), pushed close against the base. These seedheads might be unfamiliar, perhaps, in this guise – they are usually picked when they are pale green, but they are even more striking when used as they are here.

Towards the centre of the design a deep cushion effect is created by long spikes of pink polygonum and mugwort flower buds, pushed in at right-angles to the base. The stems, left at about 4 in (10 cm) long give depth and dimension to the whole design and obviate any possible flatness.

Sea holly gives a hint of silver at the very centre of the heart and, in this position, is the focal point of the design. The final touch, softening the outline when all the other material has been positioned, is a row of vetch pods captured at the young, green stage and pushed firmly in between the cones.

A bowl of yellow flags *see colour photograph on page 34*

There are times when one has an urge to have a few bright flowers around the house, yet a formal arrangement would be quite out of place. Perhaps you might be on a waterside holiday, borrowing a house from a friend, or otherwise in such a relaxed mood that formality is definitely not the order of the day.

The group shown opposite, a combination of bright yellow flags and various rushes, is designed for just such occasions. The container is a glazed pottery country casserole decorated, as it happens, with a raised pattern of game birds and ears of corn. You would use whatever you could find in the cupboard that was deep and heavy enough to take the thick stems.

The pot is filled with water and has a piece of crumpled wire mesh wired over the top. Soaked foam holding material would be better, but who has a large block of that to hand on a carefree holiday? If you do use foam, take care to cut the stems of the flags at the sharpest possible angle to help them push easily through the block.

Cut the flags to a variety of lengths, essentially long, medium and short, and arrange them to create a reasonable balance. Provide a background for them with sprays of grasses and rushes of contrasting types, some long and thin and some wide and fluffy. Soften the rim of the container by placing a few sprays of leaves – beech, lime or whatever – to fall over the front.

These flowers have large appetites for water and the container or the foam will need constantly topping up to make them last longest indoors. Keep the group in a reasonably cool place if you can. We set it against an old rush-cutter's board on a strip of rush floor matting. A casual, nearly natural arrangement of this kind takes well to wood, bricks or matting.

The black and white photograph shows the arrangement where it was worked – against a pyramid of rushes drying on the banks.

Sunburst of lilies

Sometimes the nicest things happen by accident. No one would suggest cutting a bowl of elegant, wax-like water lilies for a short-term decoration indoors; surely they are at their best floating serenely on the ponds where they grow. But weighing up the pros and cons we decided to cut some to dry in silica gel crystals (see *Chapter 3*) so that we could use them in designs for years to come.

The photograph below is one of those happy chances. We floated the lilies in the largest bowl we could find to give them a reviving drink before processing. Along came the photographer and showed us that we could have the best of both worlds – a stunning display in water for a few hours and the delightful anticipation of the dried flowers, less lustrous but still ethereally beautiful, ever after.

Maritime collection

As any collector of anything knows, one of the joys of the hobby – one of the problems, sometimes – is finding a way to display the precious treasure to maximum advantage. Stamps, photographs and postcards can be stuck in albums but are not exactly decorative. And then there is the constant decision, to show them or not to show them to visiting friends.

A collector of maritime prints had an idea; she grouped together eight of her favourite cards and embellished them with the bits and pieces – flowers, grasses and leaves – she had collected on the seashore. You could copy the design with photographs, sketches,

watercolours, theatre programmes – anything that benefits from being seen as part of a unified grouping rather than singly or in a heap in an old cardboard box.

From the top left, clockwise, the prints and decorations are these: Top: 'An English flagship before the wind with other shipping', by Charles Brooking, painted about 1754; below, 'A trading brig entering the Bristol Avon', by Joseph Walter, 1838. The decoration is symmetrical. Dried sea holly leaves, now almost greyish, are used in pairs above and below the prints, with dried white xeranthemum flowers, face downwards, between them. At the top corners, pairs of small ginger-brown cones enclose clusters of florets of white statice, which is used in single florets along the sides. Mauve statice in clusters at the top centre is balanced by two deeper mauve preserved columbine flowers below.

Top right: Top: 'Victory at sea, 1793', by Monamy Swaine; below, 'H.M. Frigate *Triton*', by Nicholas Pocock, 1797. Long lime green strips of dried grass pick up the colour of the swirling sea and edge the two prints along the sides. At the top, two dried Christmas rose buds with leaves enclose a small mauve columbine flower, with white and mauve statice at the corners. Below the prints, a deep pink columbine shows how successful the crystal-drying method can be. Shown full-face, it looks convincingly 'real'.

Bottom right: Top: 'The *Britannia* entering Portsmouth, 1835', by George Chambers; below: 'H.M.S. *Victory* at anchor off St. Helens, Isle of Wight, 4th December 1805', by J. W. Carmichael. There's an exuberance in the decoration chosen for this pair of prints, with deep golden yellow dried narcissus taking pride of place. The colour is echoed in yellow buds, also dried, of common agrimony forming an arc at the top and bottom. Completing those two motifs there are small meadow thistles, snippets of cow parsley and the deep blue of viper's bugloss. The vertical decoration, dramatic against the dark green mount, is formed of lino grass, the tiny stalks eased into gentle curves to complement the rounded corners of the prints.

Bottom left: Top: 'An Indiaman in a fresh breeze', by Charles Brooking; below: 'The Battle of Quiberon Bay, 22nd November 1759', by Dominic Serres. The full drama of these moments at sea is heightened by the restraint of the surrounding decorations – all the material is silver, grey or white. Sprays of sea holly mark the corners but for the most part the leaves are used singly, in swirling wave formation. The flowers are white xeranthemum and statice.

How to make the designs

Having seen how effectively dried flowers and leaves can decorate these small-scale works of art, here are a few general rules to follow. It is important to remember that it is the print and not the flower design that is the star of the show. Choose natural materials in colours which complement the prints, either slightly deeper or slightly paler in tone and never – the worst example – a bright red flower with a pale grey print. All you would notice would be the flower.

The same care applies to the choice of the background. We used dark green artists' mounting board; this material is available in a wide range of strong and pale colours but again it is usually advisable to avoid the very brightest.

For light and medium-weight natural materials a rubber-based adhesive is strong enough, but for heavier pieces, large cones or some nuts, for example, you might prefer to use one of the clear-drying do-it-yourself adhesives such as Copydex, Elmer's or Sobo.

We had the board professionally cut and framed before working the designs. This way you avoid the breathtaking worry of having your floral decoration spoilt when the prints are away in the shop. Measure the mounts with the greatest care so that all the prints are glued at the same distance away from the sides of the frames and, as always in this kind of design, practice the arrangement of the natural materials first so that you get it right before sticking it down.

Opposite, pressed flowers and leaves, a pan of wax and plain candles show a simple idea in the making. Details on page 79

Two ways with lupins *see colour photographs on pages 44 and 53*
Few flowers can bring a breath of the sea-shore indoors as effectively as a huge bunch of tree lupins. Their scent is a blend of the essence of pollen and the very sea air itself.

The flowers have minds of their own, the stems drooping, swooping and curving every whichway, and in designs they should be allowed to follow their own natural instincts. You can copy these designs with cultivated lupins of any kind.

First, on *page 44*, the flowers are grouped in an old black iron charcoal burner. This is the kind of appliance you see outside fishermen's cottages in southern Spain and Portugal, when the tantalizing smell of grilling sardines follows you down the street. Any slightly rustic-looking metal container would do; look in the shed for an old iron cauldron, a deep saucepan or preserving pan or even a bucket. Our container has a kind of pedestal; if yours hasn't, create an artificial one by supporting an inner container near the top of the vessel.

We stood a glass baking dish inside the burner and filled it with a large block of pre-soaked foam holding material covered with crumpled wire mesh and tied in with string. The criss-cross grill went back on top of all that, so the mechanics were as steadfast as could be.

Cut the lupin stems slantwise before arranging them, for two reasons. One, so that they will have the largest exposed area through which to take up water. And secondly, so that they will slide most easily into the foam without breaking it up. These are thirsty flowers, so keep the foam well topped up with water every day.

Here there is no pretence whatsoever at a formal arrangement. The sprays curve round into an almost complete circle, and always in the way that seems most natural. It takes only a little imagination to make the design depict a catherine wheel or tongues of flame licking round the grill.

The other lupin arrangement has flame connotations, too. It is shown on *page 53* in a wide old fireplace – against black again – an iron fireback depicting General Fairfax, the 17th century soldier. The arrangement stands on a base of natural slate slab.

What could look more right with the lupins than a support of gnarled driftwood? It might have been found on the very same beach. There are two pieces, actually, pushed together to form the shape of a figure 2, the tallest one roughly half the total height of the design.

The mechanics are similar to those in the charcoal burner – a dish with foam and wire tied firmly in, and this time secured round and under the slate base.

The tallest and widest lines are defined by flowering currant leaves, dark green just tipped with red and on dramatic red stems. In front, sprays of japonica leaves spill out beyond the slate, and at one side wisps of green barley hang their heads almost horizontally.

The lupins are positioned so that they totally relate to the driftwood, falling over, round, behind and in front of it just as if they had grown that way.

The long, fine stems are Italian rye grass and the thick tufts, reminiscent of the construction of the lupin spikes themselves, cocksfoot grass. Select any grasses from your collection that will give a similar effect. The dramatic centrepiece, like a minaret, is an undeveloped head of wild onion.

Mississippi steamboats *see colour photograph on page 43*
'Midnight race on the Mississippi' is the title of this charming coloured print, with

Opposite, deep rich tones contrasting with white show how effective these candle decorations can be. Details on page 81

Danna and Fulton the names of the chugging competitors. Here is a print richly coloured, from the deep blue of the water, the mysterious darkness of the foreground trees, the vibrant coral of the bands of smoke – you can almost *hear* the sparks flying – and the brilliant shafts of light.

These colours have been matched exactly in the surrounding mount and in the flower, shell and leaf motifs which decorate it. The design is as simple as can be, chosen to emphasize the strength and vigour of the print and in no way to detract from it.

The mount is $22\frac{1}{2} \times 27\frac{1}{2}$ in (66×79 cm) and 4 in (10 cm) deep. The motifs are stylized, prim almost, each one a little statement in symmetry. The flowers, all dried in crystals, are yellow and white daisy-like chamomile, and wild rose buds (the fully opened flowers do not dry successfully in this way), with dried briar leaves.

Above the print, a deep coral and cream-coloured scallop shell forms a fan-shaped background for the group of flowers and leaves, and this motif is echoed in a smaller way at the corners, with thumbnail-sized scallop shells.

At each side, the scallops are winged by two white piddock shells, and in the lower centre, directly in line with the powerful shaft of moonlight, pride of place is given to a voluptuous heart cockle.

As always, it is essential to measure the area of the design accurately and to test position the materials before attempting to glue them. If you put a dab of colourless adhesive only at the centres of the flowers and only sparsely on the sprays of leaves, they will slightly stand away from the card and throw interesting shadows which enhance the effect.

Chapter 7

In a Golden Mood

THERE is a kind of urgency about the months between summer and winter, an awareness of the need to make the most of everything the countryside has to offer before it is snatched away for yet another year. It is almost as if, like some actors, woods and hedges put on the brightest and most colourful show of the year out of a desire to retire at the most memorable peak of their success. For few sights can be more breathtaking than the fire-like colours of the beech foliage, or the legendary glory of the maples of North America in the Fall.

This is the time of year to gather up the golden and brown leaves as they float down from the deciduous trees, and press them under a carpet for a week or two until they are dry. They will be much the same colours as those which were cut earlier in the year from the same trees, and coaxed into a nutty brownness by being preserved in glycerine; but pressed leaves will be less supple.

It is a time to marvel at the constantly changing face of nature. Areas that once, in the spring, were white with blossom, then pastel-tinted by rose briars and bramble flowers, will now be a study in red and green. For nature never leaves the countryside completely barren – the half evergreens, like privet and bramble, the russet beech leaves that cling all winter through, will see to that.

Then there are the vibrant berry fruits which, luckily, are not confined to any one type of terrain; practically wherever we live we will be able to find some, native to that particular area, to preserve in glycerine for decorations or make into jams or conserves for winter parties.

There are the orangey-red berries of the low-growing cloudberry, tightly clustered like a handful of Christmas baubles; the scarlet, acid fruits of cowberry (whortleberry) or the fruits of the cranberry. More widely found are the rose hips, often a blaze of colour long after the leaves have left the bushes; the smaller black-tipped haws, fruit of the hawthorn, and the dense clumps of the rowan (mountain ash), all three inedible in their fresh state, but invaluable as the principal ingredient of hedgerow and country preserves.

Best known and perhaps best loved of all are the blackberries, whose large, glistening fruits give us an excuse to wander slowly but not aimlessly along the hedges or make our way purposefully across the fields. Although blackberrying is such fun, such a shouting-for-joy kind of pastime, it is really most rewarding on one's own, each person taking charge of a separate patch. For this is the way to encourage wild things, all the field and woodland animals and birds, to grow accustomed to one's presence and go about their business as if no one were there.

Other black berries which are tempting, too, include drooping clusters of elderberries, like heavily beaded Victorian jet jewellery; bilberries, found in sheltered places and on steep hills and mountains, and sloes, the acid tasting blackthorn fruits which make such excellent wine.

When so much about us is preparing for winter we should be grateful for the evergreens – the holly and the ivy; juniper and yew, and all our native pines. For now we can cut sprays from these trees and plants and preserve them in the glycerine which, earlier in the year, worked its magic on other fresh green leaves.

And with summer pastimes stored away until another year, it is a marvellous time to begin making designs with all the materials we have collected, dried and preserved, to remind us for months to come of this golden season in the countryside.

Nuts and berry picture *see colour photograph on page 55*

How often have we picked up a handful, then a pocketful, and finally a basketful of cones, nuts and berries – and then wondered what to do with them? One delightful way is to make them into a three-dimensional picture which can be a study in shapes and textures and will be practically everlasting.

This unusual design shows how well the natural, rustic browns are set off by a slightly surprising colour for the background – a deep, sharp pink. Other suitable background colours would be all the acidy blues, right through to turquoise, Chinese yellow and citrus green. But for the most harmonious effect, it is best to choose a fabric which relates to one of the natural materials in the picture. It is always pleasing to have this link, however subtle it may be. In this case, it is the colour of the sea pink flowers, nestling far down in the domed, dried flower heads; this and the pink background are almost identical.

The first step is to cover a piece of stiff card or hardboard cut to fit the frame exactly. We chose a dress-weight cotton material, cut it slightly larger than the card – about 2 in more all round – and pulled it taut across the surface by the method described on *page 52*. The frame is a slim strip of gilded wood $18\frac{1}{2}$ in (46 cm) square.

The success of a geometric design of this kind depends on the accuracy of the measurements and straightness of the lines, so it must be worked out carefully first. To do this, take a piece of paper the same size as the background, draw a circle representing the

1 Lime fruit
2 Acorn cups with hawthorn berries
3 Larkspur pods
4 Field penny-cress seedheads
5 Acorns
6 Beech mast
7 Cones
8 Cones
9 Teasel
10 Dried wheat
11 Sea pink flowers
12 Hawthorn berries

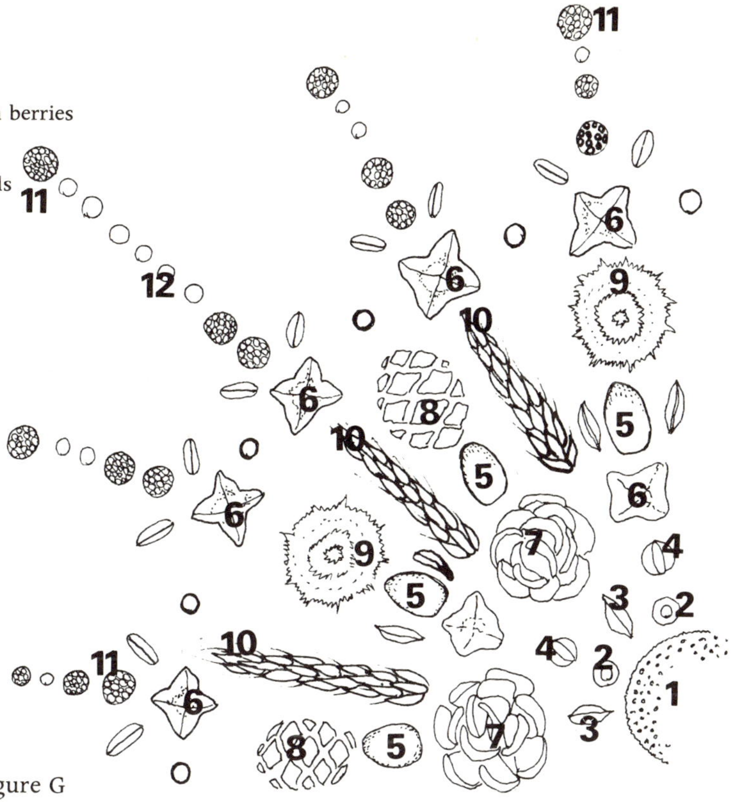

Figure G

extent of the lines, then measure the circle into equal segments. This is easiest to do, of course, if you have a set-square (try-square) and compass, but you can draw a circle with a large dinner plate or tin lid. To divide up the circle, cut another the same size in thin paper, fold it in half, then into the required number of segments – in our picture, there are ten in each half-circle.

The natural materials are a spiky, prickly lime tree (linden) fruit in the centre, acorns and acorn cups, dried teasel heads, two kinds of fir cones, beech mast, the creamy seedheads of field penny-cress and larkspur and dried sea pink flowers. *Figure G* shows a section of the design in detail.

Since so many of the materials present difficult surfaces – either very shiny or very spiky – it is essential to use a clear, strong adhesive. As you work the design, you will probably find that it is necessary to hold some pieces down for a moment or two, until the glue starts to set, or to weight them down with a piece of stick, a spoon or something similar.

Mark the centre point of the background and work outwards from there. This means positioning the linden fruit first. Then work along one row at a time, from the centre to the circumference, carefully and accurately following your drawn diagram.

If your materials are different from the ones shown, try to build up a pattern which alternates smooth and shiny textures, light and dark colours, round and oval shapes. Notice how well the pine cones and acorns relate in shape but contrast in texture; the spikiness of the lime is picked up by the prickly teasel heads; the creaminess of the field penny-cress seedheads is repeated in the much longer larkspur ones; and how the tiny acorn cups act as containers for the preserved hawthorn berries.

Dried flower arc

If your preference is for softer outlines and a more natural look, you might prefer the picture shown in the black and white photograph (*above*). The technique is just the same as that described for the nuts and berry picture, but the mood is completely different.

We had the frame first – always an advantage in this kind of work – and created a design which would be in keeping with the sturdy elegance of the gilded wood. The background is a piece of deep sky-blue cotton, the colour against which, in an ideal world, flowers and grasses are seen in summer.

First of all cut the thick card mount exactly to fit the inside measurement of the picture. Cut the fabric 2 in larger all round, then cover the card in the way described on *page 52*. Now when you have completed the design you will just have to slip it into the frame and there is no danger of its getting damaged.

Unless you are experienced in the art, it is best to choose a plain background and, as already mentioned, to relate it to at least one of the natural materials in the design. Patterned fabrics, even fussy, spriggy cottons, can make charming backgrounds but it is not always easy to visualize the relationship between background and design and how the materials will 'stand away'. Plain neutral coloured backgrounds, too, are more difficult for a beginner to handle. They give opportunities for 'tone on tone' effects with materials only slightly darker or lighter than the mount and can be stunning. But, at the worst, you can lose the flowers completely against such a close colour match.

Select the flowers and leaves for the design with as much care as you would for a flower arrangement. More, in fact, since your picture may be everlasting. For the prettiest and most easy-on-the-eye effect, contrast round shapes with flat ones, spiky and furry textures with glossy ones and, overall, aim to create a smooth, graceful outline.

Every picture needs a focal point and in this one it is the dish-shaped group of three thistles. It is as if this were a flower container and all the stalks were arranged in it. To achieve this effect it will probably be necessary to coax some stems into gentle curves, one way and the other. To do this, glue and then weight down the lower end of a stalk with a stone or pile of coins, gently bend the stalk the way you want, and glue and weight it down at the top. The merest dab of adhesive at each end and one point in the centre is usually enough.

It cannot be said too often: before starting a picture of this type, arrange your materials first and juggle them around until the shape pleases you. Follow the instructions given for the nuts and berry picture for finding the 'touch points' of the material and gluing only those surfaces. This is not a pressed flower picture, where all the materials are flat. Make a feature of this by gluing some materials at one end only so that they stand slightly away from the background and cast natural, interesting shadows. You will soon find which materials can be treated in this way and which are too wayward, and need more firm handling.

In all symmetrical designs it is important to sort out your materials and match flowers and leaves in pairs for colour, size and shape. If you fail to do this, the result will be as obvious as a strip of wallpaper hung without matching the design – and a source of irritation for just as long. If you have one absolutely fabulous flower or leaf but no partner for it, use it in a central point of the design, perhaps to cover a cluster of stalks, where it can stand proudly alone.

Follow the key (*figure H*) for the materials we used. Make a note of the different types: long or short, rough or shiny, round or flat and so on, then match type for type from the things you have in your collection. Never feel restricted by the actual flowers and leaves you see here. Whatever you use will give your picture the same overall look but it will be entirely individual.

We picked up the blue of the background with florets and bracts of hydrangea, several shades lighter. These are arranged in two curving lines on each side of the central thistles, large ones graduating evenly to small ones at the top. In deference to this really rather lovely gilded frame we kept the other materials in the beige, cream, golden colour range, deepening to pale orange in the central helichrysum flowers. (These, incidentally, were examples of flowers in our collection which had no exact equals for size and colour.)

Working order

In our picture, these were the materials used.

1 Three dried thistle heads. Position these on the card first, to find the exact area they

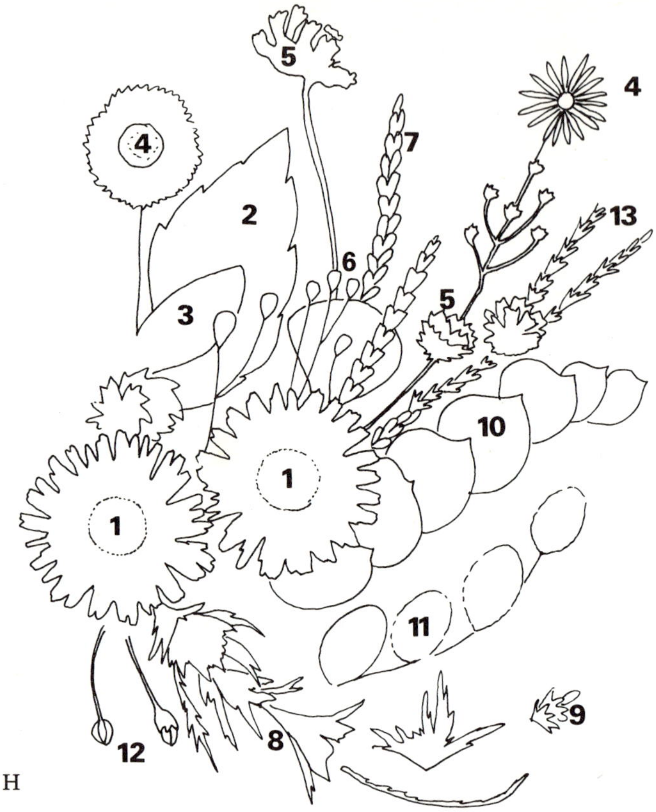

Figure H

will cover, but do not fix them in place until some of the surrounding stalks are positioned.

2 Three pairs, small, medium and large, of laurel leaves. These were preserved in glycerine until they turned a strong, shiny black.
3 Pale green sycamore keys. Position them when the laurel has stuck.
4 Helichrysum flowers. The small one, just above the thistles, is the deepest orange.
5 Dried pinks on their natural stalks. Only the lower tip of the stalk is secured.
6 Sprays of nipplewort, again with the heads standing away from the background.
7 Ears of dried barley. Any other cereal, or a grass, would do.
8 Sprays of dried dryandra. Choose a spiky material if you can.
9 Snippings of pale yellow statice.
10 Pale blue hydrangea florets and bracts.
11 Fluffy lagurus (hare's tail grass) heads arranged along false stalks.
12 Little cuttings of lino grass, dyed black.
13 Spiky dried grass, similar in feeling to the dryandra, but softer.

Against the light *see colour photograph on page 56*

Wherever you have a chance to show a group of flowers or leaves in front of a window, you have an added bonus, for you can choose specimens knowing that they will take on an extra special quality.

Opposite, a glorious harvest of dried and preserved materials centred around a dried artichoke head. Details on page 84
Next page, crystal-drying turns water lilies and Christmas roses into delightful everlasting flowers. Details on page 85

The arrangement shown in colour is a case in point. We hung it on a hook in a room while waiting to photograph it, a copper soup ladle with a cascade of yellowing leaves; nothing more. But in front of the ancient light, the tiny window with its chippings of original stained glass, it regained the almost luminous quality of leaves in shafts of natural sunlight.

The holding material is a soaked block of foam, extending well beyond the rim of the ladle and held in place underneath by a substantial pinholder. The design is based on a vague, slightly curving sweep from top right to lower left, and this is first etched in with sprays of fading beech leaves, with more to tumble over the lower rim of the containers. To trail softly in this way, rather than stick out straight and stiff, the little stems needed gentle persuasion, stroking a few times into a gentle curve.

A few sprays of late-flowering honeysuckle are clustered in the bowl of the ladle, where they conceal the green holding material, and on each side of them are sprigs of wood spurge and cow parsley (chervil) seedheads. Long sprays of dock flower heads, dry and with a matt finish, contrast strikingly with the few pieces of hawthorn with their glistening deep green leaves and shiny berries.

Admittedly this copper ladle seems purpose made for the job, but you can copy the idea with an ordinary kitchen ladle, trailing a few leaves over the handle to soften the effect and taking care to hide any harsh glint of stainless steel or chromium plate – not quite as sympathetic to flowers and leaves as burnished copper. Alternatively, you could use a small plastic bowl, even a deep tin lid, suspended on wires like a hanging basket. The actual container is not important; it simply has to be something in which to wedge the block of foam. What is important is the effect you will achieve of leaves in a golden mood, suspended against the light.

Candle mass *see colour photographs on pages 65 and 66*

Making pictures is a classic way of enjoying a pressed flower collection, producing lasting tokens as a reward for time and patience. But what can you do with your flowers when you haven't much time to spare and are in the mood for a sense of instant achievement?

The prettiest answer to the question is to decorate candles with them. Choose the plainest, simplest candles you can find, not only for the most stunning before-and-after comparison, but because they really do give the best results. Thick, stubby white ones; fat, chunky yellow ones; tall, slender purple ones, even little round nightlights can all be transformed in minutes.

It is actually easier to work on large candles, so begin with those. You really can't go wrong; if you don't like the result you can dip the candle in hot water to remove the flowers and start again; or just burn the candle.

Sort through your collection for flowers that contrast well with your candle colour. If it is white, this means that any colour from the palest – even pale pink – is suitable, and very, very effective. All the greens, yellows, browns, mauves and oranges look fabulous on white. But white and cream flowers do not show up well enough to create a pattern, and the dark blue of, say, larkspur is if anything too much of a contrast.

For yellow candles, choose cream, white, orange or brown tones, reaching to blue for more acid contrast if you wish. Pink candles take well to all the range of blue and mauve flowers, and of course white; and mauve, purple and grey are good backgrounds for beige grasses and pink, white and cream flower colourings. The photograph on

Previous page, poppies as they are usually seen, here in a rugged piece of pottery with fresh, deep green grasses. Details on page 85

Opposite, poppies again, with dried cereals and seedheads shown in a kitchen against warm, natural wood. Details on page 86

page 65 shows all the equipment you need to make these decorations: a saucepan with a little melted wax and an old paintbox paintbrush. Use a bud of paraffin wax from a candle-making kit or the ends of some white candles (they must be white) for the wax. Melt the wax in the pan over a very low heat; use a double boiler if you have one. There is no need to heat the wax beyond melting point. Work near the stove so that you can keep returning the pan to the heat to re-melt the wax – it solidifies very quickly off the heat – or stand the pan on an electric plate-warmer on the table.

You can use single flowers or leaves, sprays, or grasses for your decorations. All you have to do is to hold them against a candle and simply paint over them with the melted wax. Use quick, positive strokes to get as fine and transparent a coating as possible. This does take a little practice so don't be discouraged if your first attempt is slightly disfigured with large blobs of wax – just scrape them off with a fingernail.

The flowers will be sealed to the candle as surely and as prettily as a design painted on china and fired in a kiln. And when you light the decorated candles, the flowers will be backlit as if a lantern were hanging behind them. There cannot be a prettier way to show off their delicate, translucent silkiness.

One of the easiest flower patterns to make is a random scattering of blossoms, like the tiny cow parsley (chervil) florets decorating the yellow candle on the left of the photograph. To do this, first snip the pressed flower heads into individual florets so that they are all ready to use. Hold them one at a time against the surface of the candle and work up and down, round and round, brushing each one in place with the wax. This design looks attractive with any regular-shaped flowers, buttercups or daisy heads, for instance.

Complete sprays of flowers and leaves are just as easy to apply.

For the next candle we chose a curving stem of greater stitchwort, with the tiny white flowers that are so fiddly to handle separately. Lay the stem in a natural curve across the candle and swish over it with wax, starting at the bottom where the stem is thickest and working upwards. Sprays of vetch, speedwell and other fine-to-medium-weight plants are equally successful, but thicker sprays such as larkspur are not. They look attractive and tempting (you can see some in the envelope in the foreground of the photograph) but their stems are too thick for this technique. They need such a deep coating of wax to fix them that the flowers are shrouded and lose most of their colour.

It is fun to sort through a collection of pressed flowers and find shapes and colours which team up well. The candle on its side in front of the copper pan has a two-petalled snipping of nasturtium in the centre and a fluttering of tiny vetch or wild pea flowers all round – almost identical in shape and shade so that the smaller flowers look like the offspring of the nasturtium.

There is, of course, no need at all to use flowers and leaves of the same family together, but it is quite satisfying to do so when it works well. The next candle, a large, fat deep golden one, has a swirl of larkspur leaf, just one, like a motif of the finest deep green lace, and three deep blue florets in an arch above it. This candle shows how you can superimpose flowers on leaves, leaves on flowers, too. Make a flower collage, in other words. A tiny sprig of leaf and bud runs up the central flower. As the wax sets immediately on impact with the cold surface of the candle, there is no waiting time. You can apply the second layer immediately.

A regular spiral of small round flowers such as daisies, buttercups or cornflowers is fascinating, and equally attractive whichever way the candle is viewed. The candle in the background, on the right, has a cascading line of bachelor's button flower heads and just two of the leaves.

To define the spiral, pin a piece of string to the top rim of the candle. Wind it round in wide or narrow sweeps, according to how close you want the pattern to be, and pin it at the bottom (*figure I*). Draw round the string line with a sharp-pointed pencil or a needle point, remove the string and then follow the etched line with flowers.

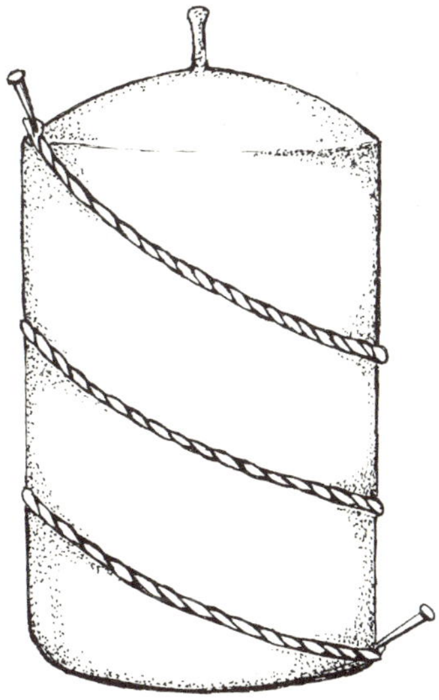

Figure I

All these designs show bright golden candles, rivalling the very flames in intensity of light and colour. The photograph on *page 66* shows colour schemes in a very different mood, reflecting the intimacy and elegance of a dinner party. The candles range from deep pink and purple to dark grey and white, each decorated with toning flowers.

First the pink candle. The deep, grainy texture of the surface, rough as leather, contrasts strikingly with the very delicate flowers, sprays of herb Robert built up into a bunch formation. A single blue flower covers the stem ends and a few tiny leaf snippings add weight and colour contrast.

The tall purple candle has the most complicated of the designs and shows the effect of a combination of leaves, grasses and flowers. First, position three delicate sprays of meadow grass, with the tallest in the centre. Then, in line with that stem, a single cow parsley (chervil) leaf, the green and purple colouring so exactly right with the candle colour. Then add four white flower buds – ours were hairy rock-cress – following the upward slant of the leaflets, with a full, four-petalled flower at the top, like a fairy on a Christmas tree. Lastly, two more flowers to flutter between the grasses – and you can almost hear the bees humming!

The candle in the centre of the photograph has a single motif. First, make a circle of separate pink and white striped rose petals, overlapping them all round. Choose one perfect round flower in a toning colour – we couldn't think of anything better than a complete wild poppy on its stem – for the bull's eye in the centre.

Tall, slim candles like the grey one could be decorated with spirals or twisting sprays of fine grasses and flowers. We decided, however, to make a virtue of the slender proportion and emphasize it with a simple stripe of cow parsley (chervil) heads and the daisy-like bachelor's button flowers. If the candle is to be used on a dining table, and will be viewed from more than one side, repeat the design on the back.

We have seen the effect of cream and white flowers on a mauve candle. Here is the colour scheme in the negative form, as it were, a single, perfect poppy dramatically silhouetted against a stark white candle. When decorating this candle we tried adding scatterings of other mauve flowers, tiny buds and little open-faced ones; arches of leaves, and feathery fronds of grasses, but scratched them all off in favour of simplicity. They detracted from the beauty of the poppy flower and contributed nothing in compensation.

If there are any rules at all, the best advice must be not to overdo the design in this kind of candle decorating. The translucent beauty of the pressed flowers calls for restraint. The effect of these candles can be ruined by over-presentation, too. They do not need elaborate and fancy candle-holders or fussy stands. Be content to stand them just on a wooden board, a dark dish, a piece of slate. Or use the plainest wooden coasters for smaller candles, such as nightlights, putting one at each place setting round the table.

It is quite exciting to see how beautifully any candles, even these tiny nightlights, respond to the flower treatment. Look for flowers and leaves in scale with them, or snip off florets and petals to use separately. You can put pairs of leaves and a single flower, to look like doll's-house flowers growing in a tub; or, much more fun, arrange the flowers to appear to hang upside-down like lanterns; contrast strong colours such as the bright yellow of a hawkweed flower with purple cow parsley (chervil) leaflets; team a pair of mauve pyrenean cranesbill flowers with a pale green leaf snipping. Or, in the kind of idiom that appeals to children, make easy-as-pie rings of buttercups or daisies, or buttercups *and* daisies, petals and leaves, all round the edge. The photograph on *page 65* shows three of these designs.

The cover design

We chose all the bright, happy, sunny golden colours for our cover design, a view of which you see in the black and white photograph opposite. The main arrangement, in an old copper mug, shows the colour strength it is possible to obtain within a very limited range – here, from the deep orange of the nasturtium to the pale creamy-yellow sprays of faded japonica leaves.

The stems are secured in a large block of soaked foam holding material, pushed well into the rim of the container and extended well above it. Needless to say, all the stems were first stood in cool water for a long drink, and the woody ones were lightly crushed to help absorption.

In a design of this kind it is advisable to strike a good balance between the two main flower shapes, the trumpet-like nasturtium and freesia, and the flat faces of the single chrysanthemum and gerbera. First define the height and width with some sprays of leaves, then place a line of the flat flowers through the centre of the design and some

at different levels on either side. Next, intersperse the trumpet flowers, nasturtium and freesia, alternating yellow and orange. Fill in the spaces with more leaf sprays, arranging dark leaves next to light flowers and vice versa. For a complete change of shape and texture, there are three sprays of under-ripe elderberries. It isn't every day that you can lay your hands on these, but you can copy the effect they create with sprays of rue seedheads, cow parsley (chervil) flowers or seedheads and so on.

The little group is made in a deep oyster shell with Plasticine as the holding material. The yellow of the design is echoed in dried achillea, still rivalling the neighbouring fresh flowers for colour intensity.

Stage 1: outline the height and width with sprigs of leaves

Stage 2: place the flat-faced flowers in a line through the centre and at each side

Figure J

The shape is a simple diagonal, low on the left and extending upwards on the right. The achillea is snipped into florets and then regrouped into clusters. These are alternated along the line with pink-brown dried pinks, almost cone-like now, and snippings of pale mauve statice. At each end there are snippings of dried larkspur seedpods, opened out and curving this way and that; silhouetted against the dark oak polished table, low at the front, there are tiny sprays of dried fern leaves.

On the right there is a candle design decorated this time almost exclusively with pressed leaves. Deep russet maple leaves are used in pairs at the centre of the design and singly at the sides and a line of buttercup flowers and leaves marches straight up the middle. On the table, musk flowers and pale nasturtium petals were rejected in favour of the stronger colours.

Harvest festival *see colour photograph on page 75*

From the glowing effects of fallen leaves and shimmering candles, to arrangements which draw on your collection of dried and preserved materials; designs which can take their place in the home for weeks and months to come.

First, one which is almost in monotone, like a faded sepia print, the display in a beautiful old pink lustre sugar bowl. The bowl has a very wide aperture, and needs packing tightly with holding material. Fill it with foam and give it a thick collar of Plasticine (non-hardening modelling clay). This will give a firm hold to all the most flimsy stalks.

The central feature and focal point is, clearly, the faded golden head of the globe artichoke. Cut this stem to the length required, slanting it at a sharp angle so that it pushes in easily without breaking up the foam.

The round, bushy spikiness of the vegetable is echoed by the dried teasel heads. Cut these, some long and some short, slanting the stems as before, and position them next.

You will obviously use whatever you have in your collection to make up the weight of the design. At the back of the photograph, softly defining the near circular shape, there is a fan of cloudy dried sheep's sorrel, and the even cloudier feathery stipa grass. Single stems are so fragile that several were wired together in bunches and pushed into the clay collar. Other dried materials, clustered round the artichoke, are seedheads of wild clematis (traveller's joy), like huge soft balls of cotton wool, groups of *Briza max*, the soft, creamy white grass and dried giant cow parsley flowers. Any of the large umbellifers would do instead.

For contrast in colour and texture, there are sprays of blackberry fruits and leaves, clusters of hawthorn berries, and stems of flowering cherry leaves, all preserved in glycerine.

Silver sauce-boat *see colour photograph on page 76*
This arrangement shows how easy it is to combine materials of different seasons, preserved in different ways.

There is a block of dry Oasis or Fil-fast pushed well into the sauce-boat and curving well above it, and a collar of non-hardening modelling clay round the rim. This provides a firm anchor for the short stems that are required to slope away from the container at an awkward angle. The heart of the arrangement is a cluster of flowers dried in crystals: Christmas roses – another hellebore would give much the same effect – yellow water lilies and a pink columbine.

The long, fine yellow spikes, defining the height and width, were also crystal dried. They are yellow dead-nettle type of flowers, seen here with the leaves stripped off to highlight the pretty little orchid-like blooms.

Curving sprays of hang-dried golden rod, greenish hands of slender-leaved pondweed in bud, and the fluffy, floppy stipa grass complete the circumference.

Wild clematis dried at two stages, one in neat little balls and one in a furry mass, provides the centre softness against which the dried seedheads and acorn cups, glycerine preserved on the stems, are seen to advantage. From pale green to soft apricot, the dried Chinese lanterns draw together all the colours of the group on a single stem.

Wild poppies *see colour photographs on pages 77 and 78*
To some people there is no more lovely sight in the world than poppies growing wild among soft green grasses or the mellow colours of ripe wheat; visions easy enough to recreate in arrangements.

In the colour photograph on *page 77*, the design shows this effect at its very simplest: a rough-textured pottery jar resplendent with grasses, blazing poppies and shafts of sunlight.

The poppies were singed at the ends as soon as they were cut, and again when they were cut to the lengths needed for the arrangement. Details of how to do this are in

Chapter 2. It was a precaution which really paid dividends, because the flowers lasted fully two weeks in water.

The grasses are cocksfoot and tall fescue with barley, oats and wheat. They were cut just before they seeded and put, fresh, into the water with the poppies. Later, when the flowers were over, the grasses and cereals were left to dry in the container where, eventually, they faded to a range of creamy-brown shades and provided a winter arrangement in their own right.

This kind of design, which is perhaps too simple to be graced by the term arrangement, is ideal for a bathroom, where it is shown, a kitchen, family room or child's bedroom. For the most natural effect, place it by a window where the light falls on it.

From the happy-go-lucky casualness of poppies in a jar to a more formal arrangement showing how different similar materials can look. The colour photograph on *page 78* shows a group of poppies, dried grasses and cereals in a farmhouse kitchen, with a pottery jelly mould as a container, a wooden pestle as an accessory and a chopping board for the stand.

Any shallow bowl, even a soup or pudding bowl, would be a suitable container. Fill it with pre-soaked Oasis or Fil-fast anchored, if the aperture is large, by a pinholder.

Define the shape with the longest materials, reaching upwards on the left of the container and slanting downwards on the right: in this case, a thick mass of dried annual beard grass and wild oats, respectively. Other oats and dried wheat heads turn the foundation shape into a long, pointed oval, which is further emphasized by the placing of the bluey-grey dried poppy seedheads.

Last come the poppies themselves. These have to be singed at the ends as usual and are placed in and among the cereals and grasses as naturally as possible. Notice that some of them are facing forwards and others to right and left. They grow like that! And this is the way to derive the maximum variety from a single type of flower. As always, the largest, most fully opened bloom is reserved for the focal point, in the centre.

When the poppies have finished, you can replace them with other fresh flowers, re-soaking the foam with more water; or you can substitute dried flowers. In this way you can keep the basic design yet present a constantly changing display.

Falling leaves

There's a reassuring firmness about pressed leaves. They are less delicate to handle than some of the flimsy flowers and so make excellent material for a picture by a not-too-experienced designer. There are full details in *Chapter 4* about pressing and storing leaves. Now here is a way to create a design which relies for its full effect on the contrast of shapes – long points, circular, oval and ivy-leaf type – in your collection. In fact in this picture the only flower is a single cow parsley (chervil) head in the centre.

The design has an under-mounting board of deep avocado green, a colour which perfectly complements the leaf tones. The border is a strip of white mounting board, 2 in (5 cm) wide, and the centre panel $11\frac{1}{4} \times 15\frac{3}{8}$ in (29 × 39 cm), with slightly rounded corners, drawn round a coin. A $\frac{1}{2}$-in (13-mm) gap between border and panel reveals the background colour. The white card, of course, comes into contact with the glass.

Opposite, with the innocence of buttercups, a honey jar of wild mustard and charlock (yellow-rocket) and trailing leaves. Details on page 95

Next page, summer bounty captured with a handful of wild strawberry flowers and fruit and wild rose blossoms. Details on page 95

CREAM

Even with leaves, it is important to use the glue sparingly. Tiny leaves like the berberis will need only the faintest touch, whereas the larger ones, like ash and the creeper, will need a dab at intervals along the central vein and at the tips.

The leaves you have in your own collection will be the ones you will use in your picture so do not try to follow the description leaf for leaf. Have fun matching your leaves

Previous page, wild roses in a silvered wine goblet with a tiny hammered silver clock as an accessory. Details on page 96
Opposite, in a Victorian mood, tiny sprays of lilac blossom and bright pink-red campion flowers. Details on page 97

1 Virginia creeper leaf
2 Maple leaf
3 Berberis leaves
4 Bugle leaves
5 Wild cherry leaves
6 Clover leaves
7 Cow parsley (chervil) flower
8 Rowan (mountain ash) leaves
9 Geranium leaves
10 Snippings of wall lettuce leaves
11 Ash leaves

Figure K

to ours by comparing them for size and shape. Match them carefully in pairs, too, when the leaves are to balance at opposite sides of the picture, and keep a good balance of colour. *Figure K* gives a complete key to the make-up of the design.

The focal feature of the design is a large, deep red Virginia creeper leaf which seemed like a gift waiting to be picked up on a city walk. This provides a contrasting background for the stripey maple leaf, pressed at the young, bright green stage. Just above it, there is a russet-coloured berberis (barberry) leaf, one of the many used, like drops of rain, as accents in the picture. More of these tiny reddish leaves, in graded sizes, outline both sides of the Virginia creeper, with two single bugle leaves at the lower corners. The bugle leaves, almost purple in the main, have delicate soft pink vignettes outlining the veins.

Above the central leaf, on either side, are three oval-shaped wild cherry leaves and a larger one, russet and nearly yellow in parts. These barely touch a three-leafed clover at each tip.

Between the two fans of cherry leaves is the only flower, the complete circle of creamy cow parsley (chervil), supporting a sturdy spray of rowan (mountain ash) leaves with, on each side, a single geranium leaf and another barberry. At the lower tip of the Virginia creeper are four wispy geranium leaves, more or less in a row, and cut-off ivy-shaped tips of three wall lettuce leaves. Pointing towards the corners of the panel, two russety ash leaves, collected from the foot of the tree at the end of the season, with again accents of barberry.

If you are to use flowers in a mainly leaf picture, take care to use them in a good proportion. Here, we show one flower used just for fun, and as an accent piece. A group of three flowers would serve the same purpose, with perhaps one or two more scattered on the design. But beyond that, a design might tend to look out of balance until you reached a one-to-two proportion, flowers to leaves. Another point to bear in mind is that the flowers you choose must be of the type that will play a secondary role to the leaves. Do not add brightly coloured flowers or petals, or those with a strong, definite shape which would overshadow the leaves. Let them have, here in their own design, their well-deserved moment of glory!

Chapter 8

The Open Country

AS we have seen, a country walk or ramble can be a positive treasure hunt at any time of the year, yielding a gloriously varied harvest of leaves and berries, fruit and flowers, some to use in the mood of the moment and some to preserve.

We can train ourselves to appreciate this natural bounty, but children seem to have an innate if subconscious affinity with it all and take it, endearingly, for granted. There are so many flowers associated with memories of childhood and leisurely country outings interrupted every few seconds for each member of the party to see and appreciate some new find. And, of course, children's books keep alive old customs, traditions and rhymes for each new generation to love all over again.

There are the simple little pink and white meadow daisies that for generations have been threaded one into the other, as circlets for tiny wrists and garlands for little girls' heads. And the seedheads of the dandelion, fondly called clocks because of their round, open faces, and trustingly referred to whenever a child out on a country picnic wanted to know the time. If it took eight earnest puffs to blow all the seeds off the stem and send them floating away to establish next year's crop of weeds, then it was eight o'clock and well past bedtime!

More romantic children put the seedheads to a different test, blowing first for 'He loves me' and next for 'He loves me not'. Let us hope it always took an uneven number of blows to send all the seeds scurrying so that the game ended on a pleasant note!

Then there are the golden yellow buttercups that children would hold under each other's chins, chanting 'Do you like butter? Do you like butter?' And if the bright little flowers obligingly reflected on to the upturned chin, as they invariably did, then the answer, not surprisingly, was yes.

Many, many handfuls of these flowers, a present for mother, must have been thrust into jars and stood on kitchen windowsills, and it is in the spirit of those clusters, innocent of any art or training, that the group shown in colour on *page 87* was born; as a tribute to those happy-go-lucky days when a clutch of buttercups was treasure as rich as a bag of coins. The container we chose is a Greek pottery jar which once held golden honey from the slopes of Mount Hymettus. It is ideal because it has such a narrow neck that no 'mechanics' are needed to keep the stems in place.

The flowers in the photograph are charlock and treacle mustard, culled after a breathtaking scramble to the top of a high and windy bank where they were growing side by side. Both members of the brassica family and closely related to the cabbage, these flowers look unpromising at first sight, to say the least. They have just a few straggly blooms at the tips of the long, watery stems, and lateral shoots bearing some even more sparse blossoms. It is only when the stems are snipped short, to no more than 6 to 8 in long, that the eye takes in the now more concentrated colour, a yellow every bit as glowing as a field of buttercups.

The arrangement lays claim to being very little more than a few stems pushed into the jar, yet as with everything that is planned to look casual, there *is* a reason behind it. The flowers are grouped together to form a ball shape, and arranged with the fully opened flowers alternating all round with the tight clusters of greeny-yellow buds. This is the way to get the most from the flowers and emphasize the contrasts in both colour and texture.

The outline of the design is softened by a few trailing stems – pick anything that will twist and droop enough: bindweed is as good as any.

And for the photographic setting, a home-baked soda-bread loaf, a pair of old wooden butter pats (or 'hands' as they are sometimes called) and a decorated block of butter to complete the farmhouse feeling. But, although you can almost imagine the golden fields stretching out from the doorstep, it is only a feeling, for the arrangement was actually photographed on the very outskirts of a capital city!

Singing of roses

'Love is like the wild rose-briar', wrote Emily Brontë, who went on to describe how its summer blossoms scent the air. She was writing about the season of awakening and exuberance, when the countryside is adorned with the soft pink of wild roses.

You can make a beautifully casual arrangement with the very smallest of wild rose posies. The design shown in colour on *page 88* uses only four, to be precise. Four wild roses, a handful of wild strawberry flowers and the nearly ripe fruit set a scene which borrows something of the feeling of an early oil painting.

The container is a tiny metal urn, with the word 'cream' helpfully engraved on a neat little brass plate. And so it is a real strawberries-and-cream-tea kind of arrangement, the sort of thing that would look pretty in a country kitchen, or, as in our setting, against the rich polished mahogany of a sailing boat galley.

To copy the design, choose a small copper or brass measuring jug, a Turkish coffee pot or a piece of unpolished pewter. Crush the rose stems lightly before putting them in water so that they will last as long as possible.

The basic shape of the design, etched in first with the roses, is an elongated S. Place the roses so that some droop well below the neck of the container. You will probably have to coax them gently into a curve with your fingers. Cluster the stems of wild strawberry around the roses, exaggerating the shape of the original outline and studiously avoiding making the design look either planned or formal.

The roses should, if you are lucky, last for a week, perhaps until the time for the next picnic comes around, and the strawberries for a little longer. Though the fruit will stay tantalizingly under-ripe!

See how much more elegant and sophisticated the flowers can look in the colour photograph on *page 89*. The container is metal again – the material seems to be a perfect complement to the shrubby briars. This time it is a Spanish silvered wine goblet with a tightly clinging vine of grapes and leaves twisting up the stem. You can use any kind of goblet on a stem, a candlestick or even a large wine glass.

To achieve the draping triangular shape of the design you will need to use Oasis or Fil-fast. Soak a fist-sized block of this material until it can absorb no more water, then push it firmly into the neck of the container to extend well above the rim. Lightly crush all the stems before arranging them.

Select two matching stems of briar, one curving to the right and one the other way, to give outline to the base. Then position the topmost stem, long and straight. You will then have the three points of the triangle.

Now choose the five largest, most fully open flowers. Cut the stems short, lightly crush them again, and position them to form a cluster in the centre. They are the focal point of the design.

Lastly, fill in both sides, keeping strictly within the imaginary lines of the triangle. Be sure to use some tightly furled buds, not only for the variation in shape, but because they will be much darker than the fully opened flowers and provide a pleasant colour contrast, too.

A pair of these goblet arrangements would look beautiful on a dinner table, especially with linen and china in soft pastel colours. But we show the flowers against a deep chive green wall in a bedroom, the background colour only just projecting the dark green leaves, and the curtain pattern a deeper tone and larger version of the flowers. The little hammered silver carriage clock repeats the metallic theme, and is used deliberately, an accessory to the design.

When May is out and lilac in bloom

The name of the hawthorn tree has for generations been invoked by nannies and grandmothers: 'Ne'er cast a clout, till May be out,' implies the retention of every layer of winter clothing, no matter how fiercely the sun shines, until the hawthorn tree is in blossom. For in England the tree which showers the countryside with its pink and white flowers, like the confetti of spring, is nicknamed May. And maytime brings the lilac, too, nodding over the fence, richly purple or snowy white.

In the colour photograph on *page 90*, a Victorian child, in the form of a tiny hand-sized plaster ornament, holds a basket of lilac blossom and red campion (catchfly) flowers. Seen in close-up, on this tiny scale, lilac blossom registers strongly as a cluster of minute four-petalled flowers. But imagine how many hundreds of blooms there must be on a large tree! The red campion flowers – not red at all, but a deep pink – are much larger than the lilac. Their long stems have been cut down to be in proportion with the little vase.

Although this pretty little-girl container plays an important part in the design, it is not essential. You can achieve a similar effect by improvising with any squat, cylindrical vase, tumbler or glass to hold the flowers. Stand a figurine, a little cherub if you have one, or even a tiny doll, beside it and bind the two together with transparent adhesive tape (Scotch tape).

You can fill the container either with soaked foam holding material or with water and a handful of small marbles. If you do not have a toy cupboard to raid, you can buy marbles in packets from most florists' shops.

Lightly crush the stems of the lilac and push them well into the container, some close against the little figure and some spilling out and away from her. Position a spray of leaves to one side – we chose the very young, almost transparent leaves of maple, and then a long, curving trail of ivy. Our leaves were fresh, but you can use preserved ones if you have them.

Lastly, position the campion, cutting some curving stems so that they drape over one shoulder of the figure, some to stand taller than she is, and others nestling close round the rim of the container.

An arrangement of this kind is perfect in a bedroom or on a lady's small writing table – somewhere where it will be surrounded by objects in the appropriate scale, yet can rightfully steal the limelight.

From lilac flowers in miniature work to a display of hawthorn blossom as we are accustomed to seeing it, in snowy clusters against its own bright green, oak-like leaves.

The design shown in colour on *page 99* is on a kitchen windowsill where it lends perspective and dimension to a limited courtyard view. When you are designing an arrangement for a particular situation like this, measure the available space – in this case the distance between the shelf and the top of the scalloped blue window blind. Condition the stems by lightly crushing them as soon as they are cut and again when you have recut them to the lengths you need. Keep them in water at all times.

The container in the photograph is a simple white coffee pot, with its lid used beside it as an accessory. At least, if you do this, you are bound to remember where you put it when you want to use the jug for coffee again!

As the pot has a fairly narrow neck, it is possible, though not always entirely successful, to make the arrangement without the help of holding material. If you wish, you can use a ball of crumpled wire mesh pushed deep into the pot to hold the base of the stems. Another way is to use short stems of evergreen, such as cypress, bunched upside down in the pot. The spreading leaves act like grabs on the stems you arrange. But evergreen leaves used in this way do tend to discolour and taint the water in the container, and many people consider that a serious disadvantage.

Cut the hawthorn stems which will form the outline of the design. Notice that they are of equal lengths, some straight for the upright line and others curving to the left and right for the side definition. When you have achieved the balance of this shape, cut shorter stems to fill in the design so that there are no gaps.

There is an added fun touch: see how amusingly the spray of oak apples round the rim of the jug echoes the shape and colour of the acorn pull on the blind cord beside it!

Three in a row

It was Mary, Mary, Quite Contrary whose garden grew 'With silver bells and cockle shells, and pretty maids all in a row'. That happy little rhyme, tinkling with the merriment of the children who perpetuate it, has a sound basic idea: the natural affinity between shells and gardens. For shells make some of the prettiest garden ornaments; they can be strung together, like a giant necklace, to edge a path or special flower bed, perhaps one set with silver-leaved plants; they can act as jardinières for a selection of tiny alpine plants (*see Chapter 9*) and make enchanting containers for little arrangements of fresh leaves and flowers (*see the colour photograph on page 100*).

This group shows three miniature arrangements with a linking theme: the only colours used are purple and green. All the plant material was gathered in a river valley, some from fields where farmers were cutting reeds for thatching houses and barns, some from boggy marshland where the collectors had to be supplied with rubber boots and some from the soft and gentle river banks which offered mooring for the boat.

The design on the left of the group is built up in a small, craggy oyster shell, the rough texture of the outside such a perfect contrast to the smooth pearly-like inner curve. Soaked holding material is used to hold the stems, but because of the wide, outward-curving nature of the shell, it needs securing first. In the base of the shell, a tiny pinholder, the smallest one can buy, is anchored with extra tacky clay substance such as OasisFix. A small clump of soaked foam, a piece with a diameter of about $2\frac{1}{2}$ in (6 cm) is pushed into the spikes of the holder.

There is a strong colour emphasis in this design, ranging from the very deep, rich bluish-purple of woody nightshade to the saturated depth of the water mint leaves. These were cut before the plant was in flower and admired not only for the strength of their colour but also for their distinctive shape, almost like little green velvet hearts.

Softly, around the edges of the shell, there are short sprays of marsh forget-me-not, the stems cut short and the long, pointed leaves discarded.

The long spikes of the pinkish-purple marsh thistle show in stark silhouette against the pale creamy background of the shell and this colour, the milky whiteness, is repeated in a few bramble flowers.

The other oyster shell design, on the right of the photograph, is a study in texture, and a deliberate tribute to all the lovely and varying shades of green we can find in a countryside of this type – and that we envy if we live in more barren terrain.

The holding material is just as described for the other shell, a thumbprint of tacky clay, a little pinholder and a block of soaked foam. The design follows a single firm, straight line, upwards to the left above the shell and downwards to the right, trailing over the map.

The outline is firmly drawn in with stems of dried rushes, cut off sharply and diagonally at the top. The stems are still host to climbing trails of convolvulus leaves, the partially opened clusters curling away at the top. All kinds of marsh rushes find a place here – the bright green triangular stems of lesser pond sedge with their long, oval spikes of dark

Opposite, on a larger scale, hawthorn and oak apples in a coffee pot make a fresh design for a kitchen. Details on page 97

Next page, a restrained green and purple colour theme links the miniatures, in shells and a pepper pot. Details on page 98

BEESTON PARK
St. Michael's Ch.
St. Lawrence's Ch.
TO LODDON 21 M.
FROM LODDON
The Grange
To Brundall
Yacht Station
Turn Right
Brundall Barn
Norwich 3
Yarmouth 19
Norwich 4
Yarmouth 18
Manor Ho.
Postwick
Postwick Grove
All Saints Ch.

I ENVY THEM NOT I DECLARE IT
MY OWN LAMB
CHICKENS AND HAM
MY OWN FLEECE AND I WEAR IT
LAWNS I HAVE BOWERS
FRUITS I HAVE FLOWERS
LARK IS MY MORNING ALARMER
SO JOLLY BOYS NOW
GOD SPEED THE PLOUGH
LONG LIFE AND SUCCESS TO THE
FARMER
PRODUCETH

green flowers; soft rush with its fluffy clumps half-way down the stem (though we cut off the stem tops) and great pond sedge which were found, gleefully, with almost L-shaped heads

The pale green clusters forming the central, focal point of the design are unopened buds of rue – any seedheads would serve the same purpose – and the prickly leaves are nettles.

At the back of the group of three designs, in soft shades of mauve and with only a touch of green, is an arrangement in an old painted pepperpot. The deepest colour is provided by a mauve orchid, just one in the centre. (It goes without saying, incidentally, that one does not pick orchids found growing wild, but can choose to do so if one has naturalized plants in one's own garden.) A dilution of that colour comes in the pink ragged robin which was carried to the boat deep in cool water, in a glass jar, to stop it wilting. The wispy outlines are like parting clouds wafting across the background; the more tightly concentrated flowers of the pink valerian give weight at the centre and on each side. A few curving stems of dried-on-the-plant rue seedheads, like miniature nuts, give texture contrast, and the fern-like leaves of bur chervil draw in a few wispy trails.

Reeds and rushes, nettles, bindweed, brambles, not the most promising of materials that grow wild in the open country perhaps. Yet in these miniature designs they prove that beauty is in the eye of the arranger!

Previous page, shells as jardinières, planted with a collection of small alpines for an indoor garden. Details on page 109
Opposite, rich, purple cooking with blackberries: lattice flan, ice cream and a soft, spicy chutney. Recipes on pages 116, 115 and 122

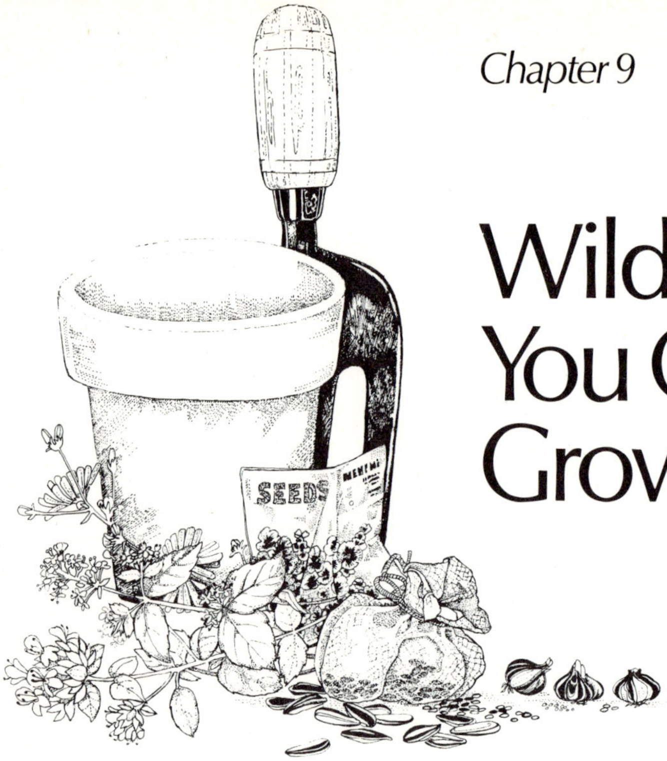

Chapter 9

Wild Flowers You Can Grow

THE random effortlessness of plants growing wild and untamed, and the always perfect relationship of size, shape and colour one to the other appear as a minor miracle to many gardeners. For, let's face it, with a little knowledge and a lot of hard work, many of us never manage to achieve anything like so perfect an effect.

If these plants can grow so luxuriantly and successfully in the wild state, surely we should be able to grow them at least as well in the garden? The answer is that we can, providing the conditions are right. This means that both the soil type and the climate must be reasonably comparable to that of the plant's natural habitat. In other words, if you want to introduce flowers which grow wild near your home, or in similar conditions, your success rate is likely to be high. But if you take a fancy to a collection of plants you see on a dry, arid patch and your garden happens to suffer from drainage deficiency, you are probably in for a disappointment.

There are three ways of achieving a wild garden. You can, if you are lucky, take one over. This might be the case when a house is built on a piece of waste ground, maybe a plot where a large old house was demolished and the garden has been running wild for years. Or you might buy a building plot in a meadow or woodland and have the joy and excitement of seeing the wild garden flourish anew with the coming of each season. In these circumstances, it is best to be patient, wait to see what grows and make notes as each patch of bulbs or shrub comes into flower. You will soon be able to create a controlled wild garden by encouraging the plants you like and discouraging the ones you don't.

The second way to attain a wild garden is to let it 'just happen' as it did to my husband and me a couple of summers ago. It was the year when we had no help at all in the garden and scarcely found time to mow the lawn. One of the herbaceous borders which had been our special pride and joy and which we had weeded diligently every week got completely out of hand. In the space of what seemed only a very few weeks self-sown wild flowers and – frankly – weeds began to flourish there and were soon in total command. The result was rather surprising. At length, and grudgingly, we had to admit that we actually preferred our new wild border. It was a mass of harmonizing pinks, reds and blues which, strangely enough, had sown itself in an orderly manner, according to the best traditions of garden planning, with the tallest material at the back and the shortest at the front.

The third way to create a wild garden is to plant it from seed. And what a sense of achievement when seeds you have collected from plants growing wild begin to show their heads in your garden or window box! Part of the fun of wild gardening is in watching for the right time to harvest seeds, when they are dry and beginning to drop from the plants. But if this is not possible and you have to collect the seed before it is ready, you can dry it slowly, and quite successfully, on trays in the sun (as long as there is not a breath of wind!) or in a warm place indoors. If you cannot collect wild seeds at all, you can buy a large selection from many professional seedsmen, together with full sowing and cultivation instructions.

There is one way *not* to obtain a wild garden, and that is by digging up and transplanting wild plants. Although it might be tempting to do so, it is quite unthinkable. If everybody did it, there would be no wild plants left for the rest of us to enjoy. And it is not even practical, because plants do not take at all kindly to a sudden change in conditions. They will have been used to thriving in non-cultivated soil with all its drawbacks; perhaps have developed root structures to cope with obstacles such as a large crop of stones or a mass of tangled tree roots. They will not be at all happy about a change to a richer diet of cultivated soil and seemingly more ideal conditions, and will probably die.

The 'take-over' garden

I have seen an inspiring example of a natural wild garden on a hillocky, hummocky piece of land, where a farm building was carefully restored to become a home again and the existing natural garden was encouraged to flourish more or less in its own way. The land was rough meadowland and, in a sense, it still is. The soil is sandy, but the owner introduced patches of chalk; this more limy soil enabled him to extend the range of wild plants that he could grow.

When he took over the property, there were about fifty different indigenous or self-sown plants. As work progressed on the house, and before he moved in, he made a note of all these, where they grew, what height they were, and what colours, and this gave him a basic garden plan. Over the years he has retained and encouraged most of the original fifty plant types, discarded only a few and introduced a number of others from seed.

A garden of this kind is no place for velvety green lawns mown in broad stripes, but rather for a friendly patch of grass to provide a background for the clumps of interest and colour growing out of it. Here and there are large patches of ox-eye daisies which spring up, come into exuberant white and yellow flower and are then cut down to grass level to lie dormant for another year. And on a smaller scale, patches of the delicate hare's-foot trefoil make smoky greyish trails in the grass. It is easy to mow or cut round them; impossible to be heartless enough to cut them down.

Splashes of colour burst forth in the form of yellow and mauve vetches, and wild violets; and a large old tree trunk, as big as a table, is a perfect foil for two more vetches, climbing purple ones. Wild thyme makes a wonderful ground cover, soft and springy to walk on, but most of all delightfully fragrant.

Viper's bugloss grows well on decidedly infertile land and appreciates the sandy soil of this wild garden, since its natural habitat is dry pastures, dunes and sea cliffs. Much the same conditions favour the tree lupin, with huge curved spikes of sulphur yellow to pale cream flowers, which grows abundantly here along a sandy bank. The shade of this same bank suits patches of bell-flowers, with their huge, trumpet-shaped flowers, which have only recently been introduced into the garden yet look as if they had belonged there for ever, emphasizing the importance of sensitive positioning when dealing with wild plants.

The owner carried out an interesting experiment with red valerian, which he grows principally to attract butterflies. He collected seed from wild plants and sowed it in two situations – in troughs against the south wall of the kitchen and in cracks in the paving stones beyond. This plant, known as a 'garden escape', which tends to become naturalized on walls, rocks and banks, flourished in the stones and did noticeably less well in the troughs.

There are other plants which this inspired gardener grows especially to delight butterflies and moths, too, such as the mauve restharrow, which he planted because moths like to breed in it; then scabious, privet, buddleia and michaelmas daisy.

The patches of chalk are a haven for bright yellow clusters of bird's-foot trefoil and the yellow gentianella, slender and upright. Wild strawberry, pretty both in flower and fruit, grows there too and behind it, still on chalk, the deep bluish-purple clustered bell-flower.

Whatever you do to create a wild garden, though, some of the plants are almost bound to prove that they are unpredictable, and will surprise you. The creator of this garden collected and planted seed from the yellow cowslip, but it produced an F1 hybrid and blossomed out in all the mixed bright colours of primula.

Around the perimeter of this carefully planned wild garden, more wild-looking than ever, waist-high grass draws a curtain over bright pink ragged robin, red campion and many colours of foxglove. Seeing them here, like this, one has the firm impression that this is how flowers were meant to grow, and this is how they like to grow.

The wild herbaceous border

My herbaceous border had always been a struggle to cultivate, because it is edged on the road side by a 12-ft-high brick wall which, unfortunately, faces north. This means that the medium loam soil beside the wall is always slightly damp, because scarcely any sun reaches it, though the front of the bed, which extends forward some 8 ft in places, is in full sun. Thus it offered a wide range of conditions to the vagrant plants and seemed to provide a perfect nursery bed for all-comers.

There are established buddleia trees at intervals along the border, again to attract butterflies. These trees became climbing posts for the pink-flowered bindweed which, although quite attractive if confined to specific areas, does need constant attention if it is not to get out of hand and suffocate less vigorous plants. Traveller's joy wound itself round the established honeysuckle and climbing roses against the wall.

The tiny white flowers of hairy rock-cress shine now like stars from the mortar between the bricks. This is a plant which takes well to wall or paving situations and multiplies quickly. It breaks up the bare areas of wall and is welcome to stay.

In the damp strips of soil beneath the wall common meadow rue now flourishes. Its natural habitat is damp meadows, ditches and stream-sides, so presumably it feels at home. The greatest advantage of this plant, to my mind, is that the dried seedheads are so useful – I have included them in many of the designs in these photographs.

Foxgloves grew and multiplied and grew and multiplied at the back of the bed in a way that they just refused to do under cultivation. By now there are clumps of two or three dozen plants in places. In front of these, perfectly matched for height and contrasting in texture, are red and white campion, the slender herb Robert with each flower head like a burning poker, and both the American willowherb and the more spiky rosebay willowherb, vibrant coral red.

The common red poppy and the long smooth-headed poppy have taken a liking to the situation and provide three quite separate sessions of colour. First, the rather lush green of the leaves and the furry buds; then the flaming colour of the flowers, and lastly the greyish green of the seedheads which dry so well. The petals of these wild poppies press beautifully and turn the colour of ermine velvet when they are dried.

Pink musk mallow and the common mallow took up a vigorous stand in the middle area of the border. These mallows are invaluable, too, for they offer virtually a second flowering season when the star-like seedheads turn silvery white and dry on the plant.

Enormous patches of pinky mauve cranesbill appeared, their flowers so soft and pretty they just asked to be pressed and used in pictures.

Deep gentian-blue larkspur has seeded itself over and over again, the deepest colour in the border and a striking contrast to many of the more delicate ones.

The pink and blue lungwort, another 'garden escape' (though it did not escape from our garden) and the bright blue alkanet grow side by side, their hairy leaves dull and greyish.

Not quite in line with the pink, blue and red colour scheme, but pretty in their own right, the yellow and white daisy-type flowers, stinking chamomile and bachelor's buttons, developed rapidly from their first appearance as seedlings to the sturdy little bushes in full flower.

A rather fragile-looking weed, the all-green sun spurge, surprised us by lasting remarkably well indoors in water. When pressed it takes on the most attractive pale and lush green shapes and patterns. Others of the euphorbia family were uninvited guests too and are now flourishing.

At the front of the bed, wild strawberries are multiplying with the clear intention of completely edging the bed in time; if they do they will meet with no opposition from us.

In fact, this wild garden which arrived like an unwanted visitor and has been more than willingly allowed to stay, is a far more 'professional' border than we had before. And it has one great advantage – it virtually never needs weeding.

Sowing the seed

Seed you collect from wild plants needs just the same care and attention as the seed you buy in a packet – the only difference is that wild seed does not come parcelled up with the relevant cultivation instructions.

First, it is important to know which group a plant belongs to. Perennials are the most hardy. They flower each year and last for an indefinite period. They include sea pink, old man's beard, water flag and tree lupin.

Biennials are sown one year to flower the next, after which they usually die. They include honesty, foxglove, teasel and evening primrose. Though you will probably still achieve a succession of the plants because they tend to re-seed themselves.

Hardy annuals are plants which will withstand frost and can be sown directly into the garden. They germinate, flower and die all in less than a year. This group includes bugloss, poppy and convolvulus.

Half-hardy and tender annuals are killed by frost; in other ways they are similar to hardy annuals. Among them are cosmos, helichrysum (straw flowers) and statice or sea lavender.

Sow the seed of perennials in a temperature of 55–60 °F (13–16 °C) in a greenhouse or cold frame. Many will flower within six months if sown early enough but others, such as some of the primula family, are slow to germinate and can lie dormant for up to six months. So do not give up hope too soon. Keep the seedlings moist and partially shaded. When they are large enough to handle, transplant them into a prepared flower bed and protect them from extreme conditions. Depending on the type, they will be ready in three to five months to plant out where they are to flower in the garden.

It is possible to plant the seed directly into a prepared seed bed, but very fine seed always benefits from being sown in a box or cold frame first.

Seed of biennial plants can be sown in a greenhouse at a temperature of about 55 °F (13 °C) or in a specially prepared bed in the garden. Box-sown seeds should be transplanted into a prepared bed when they can be handled safely, and finally into their flowering positions about two to three months later.

Hardy annual seeds should be sown where they are to flower, in soil well raked over until it is fine. The seedlings should be thinned out when they are 2–3 in (5–12 cm) and planted at a distance of three-quarters of their eventual height. For example, if a plant is likely to grow to 1 ft (30 cm) high, thin the seedlings to 9 in (23 cm) apart. Transplant the others to another part of the garden, at the same distance apart. Water them until they are well established.

Half-hardy annuals are best sown in a temperature of 60–65 °F (16–19 °C) in a suitable, well-prepared soil. Cover all but the very smallest seeds with finely sifted soil; these tiny ones can be left without covering. Seedlings should not be transplanted out into the open until there is no danger of frost.

The seed of hardy shrubs and trees should be sown in boxes or pots of rich soil, covered lightly and pressed down. They should be left in the garden (where, in some areas, the seed will benefit from being alternately frozen and thawed). After six to eight weeks it should be brought in to a temperature of 60 °F (15 °C), kept in the shade and watered so that the soil is constantly moist but never soggy. You can expect some seeds to germinate any time after the following two weeks, though some types might take as long as a year and others come through at intervals during the year.

The seedlings should be transplanted into a cold frame or prepared bed when large enough to be handled. It is important to shade these plants from direct sunlight. They should be planted out into their permanent positions the following year.

The wild garden indoors

You can bring established wild plants indoors, or plant them in shells, tubs or pots to decorate the terrace. The container should have a wide aperture and enough room inside to allow the roots to develop. It must also be strong enough to withstand some pressure when planting.

Perhaps the ideal shell for planting is the largest type in the world, the giant clam, which is found in coral reefs. These bi-valve shells can measure up to 5 ft long and each valve is so heavy that it takes two men to lift it! They are sometimes used for decorative purposes, in place of tubs or urns, in stately home gardens.

More practical are the shells we chose to photograph on *page 101*. Both of these shells belong to the triton family, which live in tropical waters near coral reefs and are usually stocked by specialist shell shops. Nearly all tritons are roomy and strong enough for planting.

We planted the larger shell with wild thyme, the small yellow-flowered potentilla verna, and a clump of compact pink stork's-bill. The smaller shell houses *Vaccaria pyramidata* (cockle), sometimes grown for medicinal purposes and which can rise to a height of 1–2 ft. In the two-handled pottery mug there is the yellow-green sedum and *Ajuga reptens* (bugle), with its mauvy-blue flowers and leaves tinged with red.

To plant containers of this kind, put a few pieces of broken earthenware pot or a few pebbles into the bottom to help drainage, then some potting soil. Take up plenty of soil around the roots of the plants, and pack them well into the shell or container aperture. Water the plants well and keep them moist, but never over water them. Remember that there are no drainage holes to allow excess water to escape.

Chapter 10

Flowers in the Kitchen

LIVING off the fat of the land, travelling Romanies used to call it, eating the fruit, flowers and berries that nature provided; and, of course, the fungus, game, birds and fish, too.

There's a whole world of freshness to be enjoyed from the countryside: flowers and stems you can crystallize for decoration, or use to flavour puddings and drinks; berries and fruit you can make into conserve, jelly, chutney, fruit butter, cheese and curd; herbs and berries to blend into clear and creamy soups, and all with such a freshness, it's like discovering a whole new way of life.

I have chosen a few recipes which make a little of the flower flavouring go a long way, or ones that combine the scarcer berries with more plentiful types, and avoided entirely any recipe, however tempting, which begins, 'take 1 qt of violet petals', or 'weigh 3 lb of cowslip flowers'.

Part of the fun of cooking is in experimenting; developing ideas and devising other recipes using different combinations of ingredients. Before attempting to use wild flowers, fruit and berries other than those mentioned here, however, be sure to check that they are edible. Some are not. As a quick reference guide, here is a list of edible, non-edible and frankly poisonous berries. Do take care that children are never allowed to go berrying unless and until they know which are which.

Opposite, storage jars, bottles, pots and an eggcup decorated with pressed flowers and grasses. Details on pages 123–125

Edible	Inedible, but safe	Poisonous
Bilberry	*too bitter to eat raw but fine for cooking*	Black and white bryony
Blackberry	Hawthorn hips	Black, deadly and woody nightshades
Blueberry	Rose hips	Guelder rose
Cloudberry	Rowan (mountain ash) berries	Holly
Cranberry	Sloes	Privet
Dewberry		Spindle tree
Elderberry		Yew
Raspberry		
Wild strawberry		

American equivalents are shown in this chapter in brackets
A metric conversion table appears on page 125

Crystallized and candied flowers

Some wild flowers can be preserved for use as cake and pudding decorations. Cowslips, daisies, heather, primroses, marshmallows, roses and violets are all edible – and delicious. You need only a few flowers or petals to give a delicate touch of luxury to a dish, and once preserved they keep indefinitely. There are two crystallizing methods, one using a heavy sugar-and-water syrup and the other gum arabic.

To make the syrup, measure 6 fl oz ($\frac{3}{4}$ cup) of water to 1 lb (2 cups) sugar into a heavy pan. Bring it rapidly to the boil and heat to 222 °F (105 °C). Drop a few flowers into the syrup and bring it to the boil again. Remove the flowers with a draining spoon. Continue to heat the sugar until it begins to crystallize on the side of the pan – when it starts to turn white. Return the flowers for 1 minute, then lift them out of the pan. Place them on a wire rack in a warm, dry room. When they are completely dry, sprinkle them with icing (confectioner's) sugar, shaking off any excess. Store the flowers on waxed paper, spread out so that they do not touch each other, in an airtight container.

For the other method, you need 1 oz gum arabic crystals and triple-strength rose water. Put the crystals into a bottle, cover with rose water and replace the screw cap. Shake the bottle vigorously at intervals during the day until the crystals have completely dissolved. Pour the solution into a bowl and dip each flower into it, or paint it on, using a fine, camel hair paintbrush. Sprinkle the coated flowers thoroughly with caster (superfine) sugar and lay them on sheets of greaseproof paper in a warm place or in the lowest setting in the oven until they are completely dry. Store the flowers in an airtight container, as above.

Candied angelica

Crystallized angelica is the perfect decoration to show off your candied flowers – it provides the colour you need for the stems and leaves.

To candy angelica – or you can treat lovage in the same way – cut the stems when they are young. Wash and trim them to equal lengths and cook them in boiling water in a covered pan until they are tender. Remove the stems from the pan with a draining spoon, strip off the outer skin, then return them to the pan. Simmer the stems gently

Opposite, non-alcoholic elderflower 'champagne', one of summer's most delicate, delicious drinks. Recipe on page 123

until they turn deep green. Remove from the pan again and spread them out on kitchen paper. Pat them dry. Weigh the stems and allow an equal quantity of granulated sugar. Spread the stems over the base of a shallow dish, sprinkle the sugar over, cover and leave for 2 days. Turn the stems and the sugar into a heavy pan, stir to mix well together and bring to the boil. Remove the stems with a draining spoon, add a further 2 oz ($\frac{1}{4}$ cup) of sugar to the syrup and bring to the boil again. Return the stems and boil again for another 5 minutes. Remove for the last time and spread the stems on a baking tray or cookie sheet. Dry in a cool oven and store in an airtight container.

Candied hazel nuts

Make a syrup with 4 oz ($\frac{1}{2}$ cup) granulated sugar to each 1 pt ($2\frac{1}{2}$ cups) water. Stir until the sugar has dissolved. Drop the shelled nuts into boiling water for a few minutes, then rub off the skins. Simmer them in the syrup for 1 hour. Remove the nuts with a draining spoon and allow them to cool. Add 1 oz (2 tablespoons) sugar to the syrup and reboil, stirring to dissolve the additional sugar. Simmer the nuts for a further hour, when they will be thoroughly coated with the candy. Remove the nuts, sprinkle them with icing or confectioner's sugar and spread them on a baking tray or cookie sheet. Dry them in an oven at the coolest setting. Wrap each nut in foil and store in an airtight container.

Marrons glacés are made in the same way. Add $\frac{1}{2}$ teaspoon vanilla essence to the syrup before boiling.

FLOWER PUDDINGS

Cowslip cream

Cowslip flowers 1 ($1\frac{1}{2}$) cup
Pinch of mace
Concentrated orange juice 1 teaspoon
Whipping cream 1 pint ($2\frac{1}{2}$ cups)
Caster (superfine) sugar 4 (5) tablespoons
Egg yolk, 1

Lightly crush the flowers on a board with a rolling-pin. Put the cream in the top of a double boiler or a basin over a pan of hot water. Add the mace, sugar, orange juice and flowers and stir well. When the mixture is thoroughly blended and hot, strain and return to a clean pan. Lightly beat the egg yolk, whisk into the cream mixture and reheat over a gentle heat, but without boiling. Stir all the time. Cool slightly, then turn into a serving bowl and chill before serving. Garnish with crystallized cowslips and strips of angelica. *Serves 4.*

Blackberry delight

Take 2 lb (4 cups) ripe blackberries. Extract the juice, either in an extractor or by pressing through a sieve covered with a layer of muslin or cheesecloth. Pour the juice into a glass serving bowl and leave it in a warm room without stirring or disturbing. After a few hours it will set to the consistency of yoghurt. Serve with lightly whipped cream or soured cream and caster or superfine sugar. *Serves 6.*

Elder flower mousse

3 eggs, separated
Powdered gelatine 1 ($1\frac{1}{3}$) tablespoon
Elder flowers, 1 large handful
Caster (superfine) sugar 4 oz ($\frac{1}{2}$ cup)
Cold water 3 (4) tablespoons
Whipping cream $\frac{1}{4}$ pint ($\frac{2}{3}$ cup)

Place the egg yolks and sugar in the top of a double boiler or a basin over a pan of hot water. Whisk lightly. Put the gelatine and water into a small basin and stand in hot water to dissolve, stirring occasionally. Add the elder flowers to the egg yolk mixture and stir, over a low heat, until the custard thickens and coats the spoon. Remove from the heat. Remove the flowers and add gelatine in a thin stream, whisking all the time. Lightly whip the cream and fold it in. Leave in a cold place until beginning to set. Whisk the egg whites until they form stiff peaks. Fold into the custard mixture. Pour into a soufflé dish or glass serving dish and decorate with elder flowers dipped first in egg white and then in caster [superfine] sugar. *Serves 4.*

Bilberry (or whortleberry) cheesecake

Butter 3 oz ($\frac{3}{4}$ stick)
Cottage cheese, sieved, 8 oz (1 cup)
Clear honey 2 ($2\frac{1}{2}$) tablespoons
Powdered gelatine 2 ($2\frac{1}{2}$) teaspoons
Berries 8 oz (1 cup)
Sugar 4 (5) tablespoons
Digestive biscuits (Graham crackers) 8 oz (2 cups)
Natural yoghurt 5 oz carton ($\frac{2}{3}$ cup)
Juice of $\frac{1}{2}$ lemon
Water 2 ($2\frac{1}{2}$) tablespoons
Water to cover
Cornflour (cornstarch) 1 oz (scant $\frac{1}{4}$ cup)

Melt butter, remove from heat and mix well with crumbled biscuits (crackers). Spoon into an 8-in flan ring springform pan on a baking tray or cookie sheet and press crumbs firmly round the sides and bottom to form a base. Leave to cool.

Combine the cottage cheese, yoghurt, honey and lemon juice. Put the gelatine and water in a cup and stand in a bowl of hot water stirring occasionally to dissolve. Stir into the cheese mixture. Pour over the crumb base and leave to set.

Put berries in a pan with just enough water to cover. Add the sugar and simmer until the fruit is tender, but not crushed. Put the cornflour into a small bowl, add a little of the fruit juice and stir to form a smooth paste. Return to the fruit in the pan and simmer, stirring, until the juice is clear and thickened. Pour the fruit on top of the cheesecake, spread evenly and leave to set. Turn out on to a serving plate when cool. *Serves 6.*

Blackberry ice cream *see colour photograph on page 102*

Blackberries 1 lb (2 cups)
Whipping cream $\frac{1}{4}$ pint ($\frac{2}{3}$ cup)
Sugar 4 oz ($\frac{1}{2}$ cup)

Wash the blackberries and hull them. Put them into a pan with the sugar and cook gently until the fruit is soft. Press the fruit through a nylon sieve. Taste the purée and add more sugar if liked. Lightly whip the cream and stir into the fruit purée. Turn into an ice tray and cover with foil. Freeze for $2\frac{1}{2}$–3 hours, stirring every hour. The ice cream will then be ready to serve, soft, creamy and delicious. *Serves 6.*

Cranberry caramels

Cranberries 12 oz (3 cups)
Whipping cream $\frac{1}{4}$ pint ($\frac{2}{3}$ cup)
Milk 2 ($2\frac{1}{2}$) tablespoons
Chopped hazel (filbert) nuts
Water to cover
Moist brown sugar 4 oz ($\frac{1}{2}$ cup)
Butter 1 ($1\frac{1}{4}$) tablespoon

Put cranberries in a pan with barely enough water to cover and cook over a gentle heat for a few minutes, until just soft but not broken. Strain and allow to cool. Put sugar, milk and butter into a thick pan over a gentle heat and stir until the sugar has dissolved. Bring to the boil and boil for 7 minutes. Beat until thick. Put cranberries in the bottom of individual serving glasses, put a layer of lightly whipped cream and then a layer of caramel. Top with a swirl of whipped cream and nuts. *Serves 4.*

Blackberry lattice flan *see colour photograph on page 102*

Shortcrust pastry to line 8-in pan
Water to cover
Cornflour (cornstarch) 1 oz ($\frac{1}{4}$ cup)
Caster (superfine) sugar to glaze
Blackberries 1 lb (2 cups)
Moist brown sugar 4 oz (scant $\frac{1}{2}$ cup)
1 egg white

Roll out the pastry and line an 8-in (20-cm) flan case (pie pan). Re-roll the trimmings and cut into strips for the lattice. Put the blackberries into a pan with the sugar and just enough water to cover. Simmer gently until the fruit is tender. Put the cornflour (cornstarch) into a small bowl or cup, add a little of the fruit juice and stir to a smooth paste. Return to the pan and stir over a low heat until the juice has thickened and cleared. Pour the fruit into the base of the flan case. Place the pastry strips on top of the fruit in a criss-cross pattern. Brush pastry with egg white and sprinkle with sugar. Bake the flan in a hot oven, 425 °F (218 °C) for 15 minutes, then reduce heat to 375 °F (190 °C) for a further 20 minutes. Cool before serving. *Serves 6.*

Wild strawberry brulée

4 egg yolks
Milk $\frac{1}{2}$ pint ($1\frac{1}{4}$ cups)
Vanilla pod
Extra sugar to sprinkle on pudding
Caster (superfine) sugar 1 ($1\frac{1}{4}$) tablespoon
Whipping cream $\frac{1}{4}$ pint ($\frac{2}{3}$ cup)
Wild strawberries $\frac{1}{2}$ lb (1 cup)

Cream egg yolks in a bowl with the sugar until pale and fluffy. Put milk, cream and vanilla pod to heat in the top of a double saucepan or a basin over a pan of hot water. Strain on to the egg yolks, stir and return to cleaned pan. Stir constantly until custard coats the spoon. Put the fruit in the bottom of an ovenproof dish, strain the custard mixture on to the fruit and leave to cool. Sprinkle the pudding with the extra sugar to cover completely to a depth of about $\frac{1}{8}$ in ($\frac{1}{4}$ cm). An hour before serving, place under a hot grill to caramellize sugar. Crack caramel with the back of a spoon before serving. *Serves 6–8.*

Blackberry soup

Blackberries 2 lb (4 cups)
Small ½ lemon, sliced
Water 2 (2½) pints
Sugar 3 oz (½ cup)
Cinnamon 1-in stick
Whipping cream

Wash and hull the fruit and put it in a pan with the lemon, water, sugar and cinnamon stick. Bring slowly to the boil over a gentle heat and simmer until the fruit is soft. Reserve a few cooked berries for garnish and put the remainder into a liquidizer. Press through a sieve. Leave to cool and serve chilled, in chilled bowls with a swirl of cream stirred into each portion. Garnish with the reserved berries. *Serves 6–8.*

Chervil soup

2 handfuls chervil leaves
Butter 1½ oz (3 tablespoons)
Flour 2 (2½) tablespoons
Chicken stock 3 (3¾) pints
2 egg yolks

Finely chop or cut the herbs. Melt the butter in a pan, add the flour and stir to form a roux. Gradually add the heated stock, stirring, and cook for 10 minutes until smooth and thickened. Add the chopped chervil, blend well and remove from the heat. Beat the egg yolks and stir them into the soup. Reheat but do not allow to boil. *Serves 8.*

Elderberry soup

Elderberries 1 lb (2 cups)
Peel of ½ lemon, shredded
Chicken stock 1½ pints (3¾ cups)
Butter 1½ oz (3 tablespoons)
Flour 1 oz (¼ cup)
Red wine 1 wineglass (½ cup)

Wash and hull the berries. Put them in a large pan with the lemon rind and stock. Bring slowly to the boil and simmer until the berries are tender. Sieve the fruit or liquidize in a blender with a little stock. Melt the butter in a pan, stir in the flour to form a roux and gradually stir in the heated stock. Add the fruit purée, stirring until smooth and thickened. Add wine and reheat. *Serves 4.*

Rose hip soup

Rose hips 8 oz (1½ cups)
Water 2 (2½) pints
2 cloves
Cinnamon 1-in stick
Flour 1 oz (¼ cup)
Butter 1 oz (¼ stick)
White wine 1 wineglass (½ cup)
Sugar 1 (1¼) tablespoon

Put the rose hips in a large pan with water, cloves and cinnamon. Simmer until the fruit is soft; remove the cloves and cinnamon stick. Put the fruit with a little of the liquid through a liquidizer, then press through a nylon sieve. Melt the butter in a large pan, add the flour and stir to form a roux. Gradually add the cooking liquid, stirring all the time, and then the fruit purée. Add the wine and sugar. Serve hot. *Serves 6–8.*

Watercress soup

Watercress 2 large bunches
Butter 1 oz ($\frac{1}{4}$ stick)
2 spring onions (scallions) chopped
Chicken stock 1 ($1\frac{1}{4}$) pint
Natural yogurt 5 oz ($\frac{2}{3}$ cup)
2 egg yolks
Salt and freshly ground black pepper
Chopped chives

Wash and pick over the watercress, discarding any faded leaves and stalks. Chop the rest, reserving a few of the best sprigs for garnish. In a large pan, melt butter and cook onions until soft but not discoloured. Add watercress and stock and bring to the boil. Cover and simmer for 20 minutes. Liquidize or rub through a sieve. Return to cleaned pan. Beat together yoghurt and egg yolks and gradually add to soup. Reheat but do not boil. Serve hot or cold, garnished with watercress sprigs and chopped chives. If liked, swirl a little single (light) cream into each portion of the cold soup.

FRUIT CHEESES, BUTTERS AND CURDS

These traditional country preserves were made when there was a generous fruit harvest. Cheeses are stored in small moulds, preferably decorative ones, and turned out whole, to be sliced and served as accompaniments to meat, poultry and game. Fruit butters and curds are poured into pots and jars and served in place of other preserves.

You can use practically any fruit: most give good, smooth results as the preserves are all sieved.

Pick over and wash the fruit, discarding any discoloured portions. There is no need to peel or core fruit such as apples, but large fruits should be roughly chopped into pieces. Put the fruit into a pan with water just to cover and simmer until it is pulpy. Put through a liquidizer (blender) if you have one and press with a wooden spoon through a nylon sieve. Measure the purée.

For fruit cheeses: Allow 14 oz–1 lb (2–$2\frac{1}{4}$ cups) sugar to each 1 lb (2 cups) fruit purée.

Fruit cheeses are not ready until the purée is so thick that you can 'cut' it with a spoon – about $\frac{3}{4}$–1 hour after adding the sugar. Choose moulds that are at least as wide at the neck as at the bottom so the cheese can easily be turned out – some yoghurt pots are ideal. Lightly brush the inside of the mould with olive oil. Pour in the fruit cheese and cover with waxed paper and transparent paper covers. Leave for 3 months to mature before serving.

For fruit butters: Allow slightly less sugar, about 8–10 oz ($1\frac{1}{4}$–$1\frac{1}{2}$ cups) to each 1 lb (2 cups) purée.

Begin to make both fruit cheeses and butters in the same way.

Wash the pan, return the purée and add the sugar (*see above*). Stir over a low heat until the sugar has dissolved. Then increase heat and simmer, stirring constantly as the purée thickens.

Fruit butters should be simmered only until they are the consistency of thick cream – about ½ hour after adding the sugar. Sterilize glass pots or jars and pour in the fruit butter. Cover with waxed paper circles and transparent paper covers. This preserve can be used immediately and, indeed, keeps for only a few weeks.

Here are some sample recipes for fruit cheeses and butters. You can adapt them for any fruit you wish to use.

Blackberry and apple cheese

Blackberries 2 lb (4 cups)
Cooking apples 1 lb
Water to cover
Sugar – see note page 118

Wash fruit. Chop apples without peeling or coring. Put fruit into a pan with the water and simmer until soft. Liquidize (blend at high speed), then press through a nylon sieve. Weigh pulp and return to cleaned pan with sugar. Boil until the right consistency is reached. Pour into sterilized moulds and cover.

You can make sloe and apple cheese in just the same way, substituting 2 lb (4 cups) sloes for the blackberries.

Cranberry cheese

Cranberries 1 lb (4 cups)
Water to cover
Sugar – see note page 118
Seedless raisins 2 oz (⅓ cup)
Hazel (filbert) nuts finely chopped 2 oz (½ cup)
1 orange thinly sliced

Simmer cranberries in water until soft. Put into a liquidizer, then press through a nylon sieve. Measure the pulp and return to cleaned pan with sugar, raisins and nuts. Bring slowly to the boil, stirring until the sugar has dissolved. Add orange slices and simmer for a further 20 minutes until a spoon drawn through the purée leaves a trail. Remove orange, spoon into oiled moulds, cover and store.

Blackberry and apple curd

Blackberries 12 oz (1½ cups)
Cooking apples peeled, cored and chopped 4 oz
Juice of 1 lemon
Unsalted (sweet) butter cut into small dice 1½ oz (3 tablespoons)
Sugar 1 lb (2¼ cups)
4 eggs

Wash berries and put with prepared apples in a thick pan over a very low heat, or in a covered fireproof dish in a warm oven. When the juice is drawn and the fruit tender, liquidize (blend) and press through a nylon sieve. Pour into top of double saucepan or a basin over a pan of boiling water. Add strained lemon juice, butter and sugar. Stir until sugar has dissolved. Beat eggs thoroughly and add to fruit. Stir until curd thickens. Pour into sterilized jars and cover.

Quince or medlar butter

Japonica fruit, quinces or medlars 3 lb
1 orange, chopped
Water to cover
Sugar – see note page 118

Wash and chop the fruit. Do not peel or core. Put into a pan with the orange and water. Simmer gently for 1 hour. Liquidize and press through a nylon sieve. Measure the pulp. Return to cleaned pan with the sugar. Stir until dissolved, then boil until the consistency of thick cream. Pour into sterilized jars and cover.

Crabapple butter

Crabapples 2 lb
Water to cover
Sugar – see note page 118

Wash crabapples and cut in quarters. There is no need to core them. Put into a pan with water and cook until soft. Liquidize, then press through a nylon sieve. Return to cleaned pan with the sugar. Cook until the consistency of thick cream. Pour into sterilized jars and cover with waxed circles and transparent covers.

JAMS, JELLIES AND CHUTNEYS

There are a number of interesting and tangy preserves to make with fruit and berries you can pick – build up a store of red-, gold-, amber- and purple-filled jars such as you will never find on the supermarket shelves.

It is important to cook fruit for jam until it is really soft; once you add the sugar the fruit toughens. Warm the sugar in a slow oven before adding it, and stir constantly over a low heat until it has dissolved. Then boil rapidly. Put a little on to a cold saucer, leave to cool and then push with your finger. If the preserve wrinkles, it is set. Pour into clean, sterilized jars and cover with waxed circles and transparent covers. To store in jars with cork lids, brush a little melted paraffin wax around the lid to improve the seal.

Cook fruit for jellies until it is pulpy; help this process along by mashing it with a wooden spoon against the sides of the pan. Strain the pulp overnight through a jelly bag – a piece of felt or several thicknesses of muslin or cheesecloth. Do not squeeze the pulp to extract the juice more quickly; this only causes it to go cloudy. Measure the juice and allow 1 lb (2 cups or 454 g) sugar to each 1 ($1\frac{1}{4}$) pt (6 dl) of juice. Put the fruit juice and sugar in a pan over a low heat and stir until the sugar has dissolved. Test for setting, pot in sterilized jars and cover.

Chutney is very simple to make, because basically all you do is to put the ingredients into a large pan, stir constantly until the sugar has dissolved, to prevent it from caramellizing, and then pour it into jars. It is ready when all the vegetables and fruit are tender, the sugar dissolved and the liquid fully absorbed. Wax circles and transparent paper covers used for jam and jelly are not suitable for chutney. Covered like this, the preserve shrinks considerably in the jars and loses flavour and appearance. Use jars with screw-top or clip-on lids.

Bilberry (or whortleberry) jam

Ripe bilberries (whortleberries) 2 lb (4 cups)
Juice of 1 lemon
Sugar $1\frac{1}{2}$ lb ($3\frac{1}{3}$ cups)

Wash fruit and drain thoroughly. Put lemon juice and fruit into preserving pan over a low heat; crush with a wooden spoon to release juice. Simmer until fruit is soft. Add warmed sugar, stir until dissolved and bring to the boil. Boil until setting point is reached. Pot in sterilized jars and cover.

Blackberry and rosehip jam

Rosehips 1 lb (3 cups)
Blackberries 3 lb (6 cups)
Water
Sugar – see method

Wash and quarter rosehips; remove seeds. Put in a bowl (*not* a metal pan) with just enough water to cover and leave for 24 hours. Hull, wash and drain blackberries and add to rosehips. Put bowl in a slow oven to extract juice. When juice has run, put fruit and juice in a pan with equal weight of sugar. Stir over low heat until sugar has dissolved, bring to boil and boil until setting point is reached. Skim if necessary. Pot and cover.

Harvest jam

Blackberries or elderberries 2 lb (4 cups)
Sloes 8 oz (1 cup)
Hawthorn or rose hips 8 oz ($1\frac{1}{2}$ cups)
Crabapples 1 lb
Hazel (filbert) nuts shelled 4 oz (1 cup)
Water $3\frac{1}{2}$ ($4\frac{3}{4}$) pints
Sugar 4 lb (9 cups)

Wash all fruit and hull berries. Chop rosehips and crabapples into quarters; remove seeds. Prick sloes. Skin and chop hazelnuts. Put all these ingredients into a pan with the water and simmer until all the fruit is soft. Add sugar, stir until dissolved, then boil until setting point is reached. Pot in sterilized jars and cover.

You can make this jam with different combinations of fruit. Blackberries, elderberries and crabapples help with setting. If you use less of these, add the juice of 1 lemon.

Spiced barberry jelly

Barberries 2 lb (5 cups)
Water 1 ($1\frac{1}{4}$) pint
Sugar – see method
Juice of $\frac{1}{2}$ lemon
Mixed (pumpkin) spice 1 teaspoon

Wash the berries and put in a pan with the water. Simmer until the fruit is soft. Strain through a jelly bag. Measure the juice and allow 1 lb ($2\frac{1}{2}$ cups) sugar to each 1 pt ($2\frac{1}{2}$ cups) juice. Put in pan with lemon juice and spice, bring to the boil, stirring, and boil until setting point is reached. Pour into sterilized pots and cover.

This jelly is a perfect accompaniment to game meats.

Crabapple jelly

Crabapples 6 lb
Water 5 ($6\frac{1}{4}$) pints
Juice of 1 lemon
Sugar – see method

Wash and quarter crabapples. Put in a pan with the water and cook until soft. Strain overnight in a jelly bag. Measure the juice and put it in a pan with the lemon juice and sugar, allowing 1 lb (2 cups) sugar to 1 ($1\frac{1}{4}$) pt juice. Stir over a low heat until the sugar has dissolved, then boil until setting point is reached. Pot and cover.

Crabapple jelly is served with poultry, game, pork and veal.

Rowanberry jelly

Rowanberries 3 lb (8 cups)
Water 2 ($2\frac{1}{2}$) pints
Juice of 1 lemon
Sugar – see method

Pick berries from stalks, wash and put into a pan with the water. Simmer until the berries are soft and strain through a jelly bag overnight. Measure juice and put in pan with the lemon juice and sugar, 1 lb (2 cups) to 1 ($1\frac{1}{4}$) pt fruit juice. Stir over a low heat until sugar has dissolved, then boil until setting point is reached. Pot and cover. Serve rowanberry jelly with poultry and game.

Wild thyme jelly

Thyme leaves 2 ($1\frac{3}{4}$) cups
White vinegar $\frac{1}{4}$ pint ($\frac{2}{3}$ cup)
Sugar $1\frac{1}{2}$ lb ($3\frac{1}{2}$ cups)
Commercial pectin $\frac{1}{2}$ bottle

Wash and finely chop or cut thyme leaves. Put in pan with vinegar and sugar. Bring to boil, stirring. Remove from heat, add pectin and stir thoroughly. Bring to boil and fast boil to setting point. Pot and cover. Serve this jelly with pork, duck or goose.

Blackberry chutney *see colour photograph on page 102*

Blackberries 3 lb (6 cups)
Cooking apples, peeled, cored and chopped 1 lb
Onions peeled and sliced 1 lb
Pickling spice $\frac{1}{2}$ oz
Coarse salt $\frac{1}{2}$ oz
Cayenne pepper $\frac{1}{2}$ teaspoon
Ground ginger 1 oz
Vinegar 1 ($1\frac{1}{4}$) pint
Moist brown sugar 1 lb ($2\frac{1}{4}$ cups)

Wash and hull the blackberries. Put in pan with apples and onions. Tie pickling spice in a piece of muslin or cheesecloth and add to pan. Add all the other ingredients except the sugar and simmer over a low heat until the fruit is soft – about 1 hour. Press through a nylon sieve to remove pips. Return to cleaned pan, add sugar and stir over a low heat until it is dissolved. Simmer until the chutney is thick. Pour into sterilized jars, cover with waxed circles and lids.

Cranberry chutney

Cranberries 1 lb (4 cups)
Cooking apples peeled, cored and chopped 1 lb
Vinegar $\frac{1}{2}$ ($\frac{2}{3}$) pint
Pickling spice $\frac{1}{2}$ oz
Seedless raisins chopped 8 oz (1 cup)
Mixed (pumpkin) spice 1 teaspoon
Salt 2 teaspoons
Sugar 8 oz (1 cup)

Wash fruit and put into a pan with half the vinegar. Tie the pickling spice in a piece of muslin or cheesecloth and add to pan with all the other ingredients except the sugar. Simmer until the fruit is soft. Add the sugar and stir over a low heat until dissolved. Simmer until the chutney is thick. Remove the spices. Pour into jars, cover with waxed circles and lids.

SUMMER DRINKS

Elderflower 'champagne' (non-alcoholic) *see the colour photograph on page 112*

4 large heads elderflowers
Sugar $1\frac{1}{2}$ lb ($3\frac{1}{3}$ cups)
Vinegar 2 ($2\frac{1}{2}$) tablespoons
Water 1 ($1\frac{1}{4}$) gallon
2 lemons

Put the flowers, sugar, vinegar and water into a large bowl. Squeeze the juice from the lemons and cut the shells into quarters. Add lemon juice and segments, cover and leave to stand for 24 hours, stirring or shaking occasionally. Strain into screw-topped bottles. Chill before serving. It's the softest, subtlest drink ever, and delicious if served with a dash of vodka or white rum.

Elderberry cordial

Elderberries $1\frac{1}{2}$ lb (3 cups)
Sugar 4 oz ($\frac{1}{2}$ cup)
Juice of 1 lemon

Strip elderberries from stalks and wash. Put the fruit in a pan with sugar and lemon juice over a very low heat. Simmer, mashing occasionally with a wooden spoon. Put through a liquidizer or rub through a nylon sieve. To serve, pour about $\frac{1}{2}$ wineglass of cordial in a tumbler, add cold water, an ice cube and a twist of lemon peel.

Wild flower appliqué *see colour photograph on page 111*

If, as they say, it is the thought that counts, there cannot possibly be a more I-thought-of-you present than a clear and shining home-made preserve presented in a jar decorated with the pressed flowers or leaves of summer. It is the very ultimate in eye appeal and good taste.

Decorating all the lovely candles shown and described in *Chapter 7*, I got completely carried away and went round the house looking for other things to 'paint' with pressed flowers. Eventually I discarded some of my early experiments, on pieces of bone china and on glass, in favour of the chunky pottery and stoneware you can see in colour. Somehow this fresh-from-the-country craft just goes better with these homely articles.

1 Stone mortar with poppy petals and cow parsley (chervil)

2 Stone pot with single poppy flower

3 Schnapps bottle with hawk's-beard flowers and hawkweed buds

4 Mustard pot with buttercups

5 Marmalade jar with grasses and poppies

6 Rum butter jar with nasturtium petals

7 Easter egg and egg cup with wild geranium (pyrenean cranesbill) flowers

Figure L

The technique is just the same as that for decorating candles. It is only the surface that is different. You hold the pressed flower, spray of grass or leaf against the surface, brush over it quickly with hot melted wax, and there you have it. A totally unique container.

Figure L is the key to the colour photograph. It shows an old stone mortar that had long ago lost its pestle and its usefulness, stone pots that once held old-fashioned marmalade or rum butter, a Schnapps bottle, a French mustard pot and a pottery egg cup.

As far as possible, I matched the flower decoration to the contents, to make the presentation set complete. One word of warning here: fill the jar first with the preserve or jelly, cover it in the usual way and when it has set, decorate the jar. If you do it the other way round the heat of the preserve will melt the wax and wilt away your design.

1 The stone mortar makes a delightfully chunky container for wild flower pot-pourri. The recipe for this is on *page 46*. The mortar is decorated with a ring of overlapping poppy petals, dried after pressing to a deep mauve colour. Covering the base of every alternate petal there is a cow parsley (chervil) head and on top of that, a bachelor's button daisy flower. This mauve and cream colour scheme tones perfectly with the flower mixture inside.

2 This stone pot holds bilberry (whortleberry) jam (*recipe on page 121*) and is decorated with a single wild poppy, complete with stem. The pink of the poppy is just one shade paler than the berry pink of the preserve inside.

3 If you make your own wine, dandelion wine perhaps, here is an ideal way to present it. The stone schnapps bottle has sprays of hawkweed flowers and buds, with extra flowers like miniature suns, and fan-shaped sprays of leafy hawkweed buds. Choose the

brighter, chunkier flowers from your collection when decorating a rough surface like this one. The delicate flower sprays would be lost against it.

4 A French mustard pot is a super container for a golden gift of crabapple butter (*recipe on page 120*). The cork is sealed around with melted paraffin wax. The flowers are a simple spray of buttercups, some fully opened and some pressed in bud, just like any freshly picked bunch.

5 An old marmalade jar has a new cargo: harvest jam (*recipe on page 121*), with the blackberries and elderberries predominating and turning the preserve a deep purple. The decoration is a screen of fine meadow grass, then five pressed poppies in varying sizes and, for a realistic effect, a few more sprigs of grass veiling the flowers.

6 The rum butter jar now holds glowing golden rowanberry jelly (*recipe on page 122*). The decoration is a false nasturtium flower, made up of six separate petals, alternately coppery bronze and pale orange. A pressed nasturtium bud on its stalk covers the centre.

7 An idea for Easter: a plain pottery egg-cup with matching egg! It takes less time to brush on a pattern of pressed geranium flowers than it does to hard-boil the egg. Any small-scale flowers look pretty – try buttercups, daisies, potentilla, violets, vetch – a different one for each member of the family. Have some eggs in reserve actually to eat for breakfast, so that the decorated ones can be kept as treasures! To keep decorated eggshells indefinitely, 'blow' the raw egg through a pinhole at one end of the shell before painting it with flowers.

True, the flower decorations on the pottery will not last for ever, but the thought will. They survive washing in cool water once or twice and anyway they did not exactly take a lifetime to do.

METRIC CONVERSION TABLES

Avoirdupois	Metric
Weights	
1 oz	28 grams (g)
4 oz	113 g
8 oz	227 g
1 lb	453 g
2.21 lb	1 kilogram (kg)
Liquid measures	
1 pint	568 millilitres (ml)
1 teaspoon	5 ml
1 tablespoon	15 ml
$1\frac{3}{4}$ pints	1 litre (l)
1 gallon	4.8 l
Measurements of length	
1 inch	2.5 centimetres (cm)
1 foot	30 cm
1 yard	91 cm

Index